A DESCRIPTION OF GREENLAND

ARCTIC REGION AND
ANTARCTICA ISSUES AND RESEARCH

Additional books and e-books in this series can be found
on Nova's website under the Series tab.

A DESCRIPTION OF GREENLAND

HANS EGEDE

NOTICE TO THE READER

Library of Congress Cataloging-in-Publication Data

ISBN: 978-1-53615-077-3

Published by Nova Science Publishers, Inc. † New York

CONTENTS

HISTORICAL INTRODUCTION

The regions in the neighbourhood of the North Pole have lately become the objects of increased curiosity; and among these regions Greenland has attracted a more than usual interest. This country was first peopled by a colony from Iceland, which occupied both the Western and Eastern parts of the Island. The first settlers in the West appear to have been destroyed by the natives, who are denominated Skrellings; and though a communication was preserved for several centuries between the Eastern coast of Greenland and some parts of the Danish territory, yet it was interrupted about the close of the fourteenth century by accumulated masses of ice, which formed an impenetrable barrier of considerable extent around the shore; and though various attempts have been made, at different times, to explore a passage through this frozen rampart, yet there is no definite account of any attempt of this kind which has hitherto been successful. May we hope that the execution of this project, which is prompted, not only by curiosity but by philanthropy, is reserved for the present era, and that it will be finally accomplished by the nautical skill and enterprise of this country!

As we possess indubitable evidence that a considerable extent of this coast was formerly occupied by a flourishing colony, and that it contained numerous villages, with a bishop's see, we cannot but be anxious to know what has been the fate of so many human beings, so long cut off from all intercourse with the more civilized world. Were they destroyed by an invasion of the natives, like their countrymen on the Western coast? or have they perished by the inclemency of the climate, and the sterility of the soil? or do they still subsist? If they subsist, it must greatly interest our curiosity to learn in what manner they have vanquished the difficulties with which they have had to contend, both from the climate and the soil, and the total privation of all articles of European manufacture. In the novel circumstances in which they have been placed, have the present race advanced or declined in the degree of culture which their forefathers possessed? What proficiency have they made? or what deterioration have they undergone? Have they remained nearly stationary at the point of civilized existence at which their ancestors were placed four centuries ago? or have they entirely degenerated into a savage race, and preserved no memory nor vestige of their original extraction from, and subsequent communication with, the continent of civilized Europe? These are certainly points of interesting research; and to which we cannot well be indifferent as Christians, or, indeed, as human beings.

In the meantime, though we cannot yet supply any particulars respecting the present state of the Eastern coast of Greenland, we think that the readers of this new edition of Egede will not be displeased with us for furnishing them with all the information which remains, respecting its past state, as well as with some historical details, which will render the present volume more complete than it would otherwise have been.

Greenland was first discovered by Eric, surnamed Rufus, or the Red, in the year 981 or 982[1]. This chieftain was of Norwegian extraction. His father had fled from Norway, and taken refuge in Iceland, in order to avoid the vengeance which menaced him, on account of a murder which he had perpetrated in his native land. Eric appears to have committed in Iceland a crime similar to that for which his father had fled from Norway. In endeavouring to escape the pursuit of justice, Eric accidentally discovered the coast which is the present object of our inquiry. He took his departure from Iceland at the port of Snæfellzness, which is situate in a Western promontory of that island. He arrived in the vicinity of a mountain called Midjokul[2]; or, as it is denominated by others, Miklajokul. Peyrere interprets this, "*le grand glaçon*," the great mountain of ice. Subsequent navigators gave it the name of Bloeserken, or Blue Smock, and others of Huidserken, or White Smock, according to the variations in the hue of the ice in different aspects and at different periods of the year.

Eric passed the first winter after his departure from Iceland in an island which he called after his own name, Ericscun, and which Torfæus places in the midst of the cultivated Eastern district. In the following spring he entered one of the bays of Eastern Greenland, to which he gave the name of Ericsfiord; and where he formed his first settlement, which he denominated Brattahlis. In the summer of the same year he explored parts of the more Western district, and gave names to many of the places which he visited[3]. He passed the following winter in the island of Ericscun; and in the succeeding summer he passed over to the main land, and proceeded along the Northern coast till he reached an immense rock, which he called Sneefiell, or the Rock of Snow. At this point he gave the name of Ravensfiord to another bay, on account of the multitudes of that ill-omened bird with which this spot abounds. Other parts of the coast derived their appellations from the names of the different adventurers who accompanied

---

[1] Torfæi *Gronlandia Antiqua Havniæ*, 1706, p. 16; see also Peyrere *Relation du Groenland*, p. 84. These two authorities are principally followed. Peyrere's work is in Recueil de Voyages au Nord, tome premier.

[2] Torfæus, p. 13, "*medias alpes.*"

[3] Torfæus, p. 14.

Eric in this expedition, as, Hergulfsness, Ketillsfiord, Solvadal, Einarsfiord, &c[4].

In the following summer Eric, having conciliated the forgiveness, or purchased the forbearance, of his enemies in Iceland, returned to that country to procure an additional supply of inhabitants for his new settlement. In order to render his proposals more attractive, he named the country for which he was endeavouring to provide colonists, Greenland, as if, compared with the rugged sterility of their native Iceland, it was a region of verdure and delight. He described it as abounding in cattle, and as rich in every species of game and fish. And as such delusive representations, when assisted by the vivid eloquence of enthusiasm, or the unhesitating assurance of effrontery, seldom fail of their effect, Eric returned to his recent acquisition with numerous ships, and a large body of settlers, from Iceland.

In less than twenty years after Eric the Red had begun to colonize Greenland, his son Leiff, who had made a voyage into Norway, renounced his Pagan errors, and received the baptismal rite. His conversion was owing to the example and the admonitions of King Olave Tryggwine, or Trugguerus[5], who had himself recently embraced the same doctrine, and had been very successful in causing it to be diffused throughout his dominions.

Leiff, having passed the winter at the court of the King of Norway, returned to Greenland, in company with a priest and some other missionaries, whom the King had commissioned to instruct Eric, and the other settlers, in the faith which Leiff had embraced. On their voyage to Greenland they met some mariners, who were floating upon a wreck in the open sea. These they took on board, and conveyed to the new settlement. Eric, at first, incensed with his son for having laid open to strangers the route to the new-discovered country, turned a deaf ear to his Christian admonitions. But the earnestness of the son, seconded by the instruction of the missionaries, at last prevailed over the insensibility of the father, who

---

[4] Ibid., p. 19.
[5] Peyrere.

submitted to the rite of baptism, when the other Greenlanders followed his example.

The Christian doctrine, which had been thus introduced, was so much approved, and so generally received, that churches were established in twelve different parts of East Greenland, and in four of the Western district. Torfæus makes the year 1000 the era of the conversion of the Greenland colonists to the Christian faith. This historian of ancient Greenland has also preserved a list of its bishops, from the year 1021 to 1406, after which period no mention is made of any subsequent episcopal appointments; and indeed the intercourse between Greenland and the native region of the first settlers appears to have been previously discontinued.

A Danish Chronicle, which M. Peyrere had consulted, refers the discovery of Greenland to a much earlier date than that which has been given upon the authority of Torfæus; and the earlier date of 770 is more likely to be true, if, as M. Peyrere mentions, there is a bull of Pope Gregory IV, in 835, relative to the propagation of the Christian faith in the North of Europe, in which Iceland and Greenland are particularly mentioned.

The Danish Chronicle, to which Peyrere appeals, states, that the Kings of Denmark, having been converted to Christianity during the empire of Louis le Debonaire, Greenland had become an object of general attention at this period.

The Danish Chronicle relates, that the first settlers in Greenland were succeeded by a numerous posterity, who penetrated farther into the country, and discovered, among the rocky heights and icy mountains, some fertile spots, which were more auspicious to pasturage and cultivation. They followed the division of Greenland which Eric had established, and called the two settlements in the East and the West, Osterbygdt and Westerbygdt.

In the Eastern district the Greenlanders erected a town, to which they gave the name of Garde, where, according to Peyrere, who refers to the Chronicle, the Norwegians established a sort of emporium for the deposit and sale of their merchandize. The town of Garde became also the residence of their bishops; and the church of St. Nicholas, the patron of sailors, which was built in the same town, became the cathedral church of the Greenlanders.

As the temporal jurisdiction in Greenland was subject to the kings of Norway, so the spiritual power of the bishops was subordinate to that of the archbishops of Drontheim; and the bishops of Greenland are said frequently to have passed over to Norway, in order to consult their ecclesiastical superior.

The Danish Chronicle, which was one of the early documents upon which Peyrere founded his narrative, relates, that an insurrection broke out in Greenland, in 1256, when the inhabitants refused any longer to submit to the tributary exactions of Magnus, King of Norway. On this occasion, Eric, King of Denmark, at the request of Magnus, who had married his niece, equipped a naval armament in order to quell the rebels, and restore the authority of his nephew. The Greenland insurgents no sooner beheld the flag of the Danish fleet approaching their coast than they were struck with a panic, and sued for peace.

This peace was ratified in the year 1261. Angrim Jonas, who records the above-mentioned transaction, gives the names of the three principal inhabitants of Greenland, who signed the treaty in Norway. "Declarantes," says Angrim, as quoted by Peyrere, "*suis factum auspiciis ut Groenlandi perpetuum tributum Norveguo denuo jurassent.*" Under their auspices the Greenlanders had been again brought to swear to pay a perpetual tribute to the Norwegian.

In composing his account of ancient Greenland, Peyrere derived his principal information from an Icelandic and a Danish Chronicle. The first was the production of Snorro Sturleson, who was a native of Iceland, and chief justiciary of that island in 1215. We are also indebted to him for the compilation of the Edda.

In the Icelandic Chronicle above-mentioned, which appears to be a tissue of different narratives, one of the chapters is entitled, a Description of Greenland, which Peyrere has copied into his account as literally as the difference of languages would admit. There is a similar description in Torfæus (p. 42, &c.), with particular but unimportant variations. Both the accounts are founded on the authority of Ivar Bert or Ivar Bevius, who had, for several years, been steward or maitre d'hotel to the Bishop of Garde, and was one of the persons who had been selected by the governor to expel the Skrellings from the Western province of Greenland or Westerbygdt, which they had invaded and depopulated.

Perhaps it will be best to insert this description of Eastern Greenland, which was the most flourishing settlement of the Norwegians in this country, as it is found in the narrative of Peyrere, and in the history of Torfæus. If the skill, the philanthropy, and the enterprize of some English navigators should ever obtain an access to this long lost settlement, and the passage should again become as safe and practicable as it was in ancient times, it will be an interesting research to compare the present state of this district with the early accounts.

The most Eastern town in Greenland, says Ivar Bert, as exhibited in the French version of Peyrere, is called Skagefiord,[6] where is an uninhabitable rock, and farther out in the sea is a shoal, which prevents ships from entering the bay, except at high water, and it is at this time, or during a violent storm, that numbers of whales and of other fish enter the bay and are caught in abundance.

As you proceed a little higher towards the East, there is a port called Funka badir, from the name of a page or missionary of St. Olave, King of Norway, who, with several other persons, suffered shipwreck at that spot.[7]

In a still higher latitude, and close to the mountains of ice, or, as Torfæus says, "*propius Alpes,*" is an island, named Roansen or Ranseya,[8] which, in early times, appears to have been celebrated for the quantity of animals, particularly of white bears, which it furnished for the chace. Torfæus says, that these white bears were not to be hunted without leave of

---

[6] Torfæus possessed three versions of this account in German, Danish, and Icelandic. His narrative runs,—The most Eastern settlement of Greenland lies under the promontory of Herjolfsness, and is called Skagefiord. Here is an uncultivated mountain, called Barrafell. At the mouth is a long sandbank (*pulvinus longus*), stretched across, so that no ships can enter, except when the water is raised to a great height by the wind and tide. At that time numbers of whales crowd into the bay. Here is a never-failing fishery, which is part of the episcopal domain.

[7] Torfæus and Peyrere.

[8] Torfæus says, "ubi publica *in sylvis* venatio." Of what trees were these woods composed? Peyrere, copying the Icelandic Chronicle, says, "Où il se fait grande chasse de toutes sortes de bestes, et entre autres de quantité d'ours blancs."

the bishop. Beyond this spot the land and ocean are said to present nothing but an accumulation of snow and ice.

To the West of Herjolfsness is Kindilfiord, or, as Peyrere spells it, Hindelfiord, which is described as a cultivated and well peopled bay. Upon the right, as you enter the bay, there is a church, called Krokskirk or Korskirk, with a monastery consecrated to St. Olave and to St. Augustin, the domain of which extends to Petersvic, where there are numerous habitations. It also possesses the territory on the opposite side of the bay.

Next to Kindilfiord is Rumpesinfiord, or Rumpeyarfiord[9], in an interior recess of which there is a convent, dedicated to St. Olave, which is proprietary of the whole district to the shore of the bay. This bay contains many holms or little islands, the property of which the monastery divides with the episcopal see. Numerous hot springs are found in these islands, of which both Peyrere and Torfæus say, that they are so hot as to be inaccessible during the winter, but that in summer the temperature is so much reduced, that they become the resort of many persons in a diversity of maladies.

Next to Rumpesinfiord, is Einarsfiord, and between them is a large mansion, named *Fos*, fit for a king or "regi competens" in the language of Torfæus.[10] Here is also a large church dedicated to St. Nicholas. As you enter Lunesfiord, to the left, there is a little promontory called Klining; and beyond it an arm of the sea, denominated Grantvich. Farther in the interior is a house,[11] named Daller, which belongs to the bishop's see. The cathedral is at the end of the bay. Here is a large wood, in which cattle are left to browse.

The whole of Lunesfiord, with the large island which is called Linseya by Torfæus, Reyatsen by Peyrere, is appropriated to the cathedral. This part abounds with rein deer, which are hunted with the consent of the bishop. The island of Reyatsen contains a species of stone or marble, out of which they cut bowls, jugs, and different kinds of culinary vessels, which possess the property of resisting the fire.

---

[9] Torfæi Hist. p. 45.
[10] Peyrere's words are, "il y a une maison royale nommée Fos."
[11] Torfæus calls it "villa magnifica."

More to the West is an island named Langent, where there are eight farms.[12] In the vicinity is Ericsfiord; and at the entrance of this arm of the sea there is an island called Herrieven, or the Harbour of the Lord, half of which belongs to the bishop's see, and the other half to the church, which is called Diurnes, which is seen on entering Ericsfiord.[13]

The country, says Peyrere, copying the Icelandic Chronicle, is unpeopled and desert between the Osterbygdt and Westerbygdt; and upon the borders of this desert there is a church, called Strosnes, which was formerly the metropolitan see, and the residence of the bishops of Greenland. The Westerbygdt is represented as occupied by the Skrellings.[14] This part of the country is described as possessing horses, oxen, sheep, goats, and other animals, but no human beings, either Christian or Pagan.

Such is the account which is given of the ancient state of Greenland by Ivar Bert, the author of the Chronicle, which is mentioned above, and in which, if there be some inaccurate representation, there is probably more truth.

Peyrere remarks, that the Icelandic Chronicle is incorrect in describing the church of Strosnes as the episcopal see, since that honour always belonged to the town of Garde. The Danish Chronicle, whilst regretting the interruption of the communication with Greenland, assures us, that, if the episcopal residence of Garde[15] were still standing and accessible, we should find a great number of documents for a complete and authentic history of Greenland.

---

[12] Torfæus says, "rusticorum villæ."

[13] Peyrere.

[14] Doctor Wormius, who was famed for his great research in Northern antiquities, told Peyrere, that these savages, the aborigines of Greenland, inhabited one side of the bay of Kindelfiord, in the Western district, and that the Norwegians had a settlement on the opposite bank. When, then, the author of the Icelandic Chronicle said, that the Skrellings possessed all the Westerbygdt, he meant only the country to the West of this bay. A small party of Norwegians, who had passed over to the Western bank, were destroyed by the Skrellings. This caused the Viceroy of Norway, who is called judge of Greenland, to dispatch a ship with a large force to revenge this affront. But the savages, at the sight of this vessel, took to flight and concealed themselves in the woods and rocks; which occasioned Ivar Bert to represent the country as destitute of inhabitants.

[15] Angrim Jonas, of Iceland, according to Peyrere, says expressly, "Fundata in Garde Episcopalis residentia in sinu Eynatsfiord Groenlandiæ Orientalis."

The Iceland Chronicle, according to Peyrere, gives a varying and inconsistent account of the fertility of Greenland. In one part it says, that the country furnishes the best corn which is to be found in any part of the world; and that the oaks are of such vast bulk, and such stately growth, that they produce acorns as large as apples. But in another passage the same Chronicle affirms, that no seed of any kind, which is sown in Greenland, will grow on account of the cold; and that the inhabitants are unacquainted with the use of bread. The latter part of this account harmonizes with that of the Danish Chronicle, which affirms, that when the country was first discovered by Eric the Red, the sterility of the soil obliged him to subsist entirely upon fish.

But in the same Danish Chronicle, which has just been mentioned, we find it asserted, that, after the death of Eric, his successors, who penetrated farther into the country, discovered some fertile spots between the mountains, and fit either for pasture or tillage. The Icelandic Chronicle contradicts itself when it says, that nothing will grow in Greenland owing to the intensity of the cold. Peyrere also remarks, that that part of Greenland, which was peopled by the Norwegians, is in the same latitude as Upland, which is the most fertile province in Sweden, and produces fine crops of grain. And the Icelandic Chronicle itself says, in another place, that the cold in Greenland is not so great as in Norway; and very good corn is grown in that country.

Greenland, says Peyrere, like other countries, which are composed of plains and mountains, exhibits great diversities of soil, and though the close approximation to the farthest North, in many situations, destroys the process of vegetation, yet there appear to be localities, which are by no means destitute of fertility. There are pastures possessing excellent herbage; and amongst the animals, which contribute to the subsistence of man, or to other uses, we find[16] sheep, oxen, horses, rein deer, stags, and hares; and of the more savage animals, we find wolves, foxes, and an abundance of white and black bears. The Icelandic Chronicle mentions beavers and martens.

---

[16] Peyrere, p. 99.

Peyrere adds,[17] that grey and white falcons abound more here than in any other part of the world. The superior excellence of these birds caused them to be formerly sent to the kings of Denmark, who made presents of them to the kings and princes in the neighbouring countries, when falconry constituted one of the amusements of the great.

The above-mentioned author, who wrote in the middle of the seventeenth century,[18] says, that in Greenland nature produces a singular phenomenon, which is described as a sort of miracle in the Icelandic Chronicle. This phenomenon is no other than what is commonly called the *Northern Lights*. These lights are mentioned as appearing more particularly about the time of the new moon; and illuminating the whole country, as much as if the moon were at the full. "The light is more bright," says Peyrere, "in proportion as the night is more dark."

The Danish Chronicle, which is quoted by Peyrere, relates, that in the year 1271 a violent hurricane from the North East drove a vast accumulation of ice upon the coast of Iceland, which was covered with so many bears and so much wood that it led to the supposition, that the territory of Greenland was extended more to the North East than had been hitherto gained. This circumstance tempted some Northern sailors to attempt the discovery, but they found nothing but ice. The kings of Norway and Denmark had long before this fitted out ships for the same purpose, but without any more success than the Icelanders had experienced. The principal incitement to these voyages was a received opinion, or traditionary report, that this country contained numerous veins of gold, of silver, and precious stones.

The Danish Chronicle pretends, that some adventurous merchants formerly amassed a large treasure by these expeditions. But regions of silver and gold have always been amongst the favourite illusions of mankind; and the imagination has revelled in visionary mines of the precious metals, not only in the South but in the North; and both at the Equator and the Pole.

---

[17] P. 99. See also Crantz, vol. i. p. 78.
[18] Peyrere's account of Greenland is dated from the Hague, 18th June, 1646.

In the time of St. Olave, King of Norway, some sailors from Friesland, incited by the thirst of gold, are said to have undertaken a voyage to the North Eastern extremity of Greenland; but, instead of returning home with mountains of wealth, they were happy to escape the fury of the winds on this rocky coast, in any miserable asylum which they could find.

The Danish Chronicle, which is a mixture of truth and fable, adds, that the Frieslanders, having made a landing upon the coast, discovered some wretched cabins just rising above the earth, around which lay heaps of gold and silver ore. Each of the sailors helped himself to as much as he could carry away. But, when they were retreating to the shore, in order to re-embark with their treasure, they saw some human forms, as ugly as devils, issuing out of their earthen huts, armed with bows and arrows, and accompanied with dogs of vast size. Before all the sailors could reach the shore some of them were seized by these frightful archers, who tore them limb from limb within sight of their companions. The Danish Chronicle adds, that this region is so rich that it is peopled only by devils.

Peyrere tells us, that one of the chapters in the Icelandic Chronicle describes the ancient route between Norway and Greenland, before the navigation was rendered impracticable by the descent of accumulated mountains of ice from a more remote point of the North. But what is mentioned concerning this route contains nothing very definite or satisfactory.

The above-mentioned Icelandic Chronicle has another chapter on the affairs of Greenland, transcribed from an old book entitled *Speculum Regale*. This chapter describes some marine monsters of enormous dimensions, which were formerly seen upon the coast of Greenland. The Norwegians called the first of these prodigies Haffstramb; and speak of it as showing itself breast high above the waves. It resembled the human form in the neck, head, visage, nose, and mouth, except that the head was more than usually elevated, and terminating in a point. It had wide shoulders, at the end of which were two stumps of arms, without any hands. The body tapered downwards, but it was never visible below the middle. It had a frozen look. The emersion of this phantasm above the waves was the signal of a hurricane.

The second monster received the appellation of Marguguer. It resembled the female form as far as the middle. It had large breasts and dishevelled hair; its stumps of arms were terminated by large hands, the fingers of which were united by a web like the toes of a goose. It has been seen holding fish in its hands, and putting them into its mouth. Its appearance always presaged some violent storm. If it turned its eye to the sailors, when it plunged into the water, it was sign, that they would not suffer shipwreck; but, if it turned its back, it was a sure omen, that they would perish in the deep.

The third phenomenon received the name of Hafgierdinguer, which was not properly a monster, but consisted of three large bodies or mountains of water, which the tempest impelled into that form; and when, unfortunately, any ships happened to become engaged in the triangular surface, which these three mountains formed, there was but little chance of their escape. This marine monster appears to have been engendered by strong currents conflicting with opposing winds, which suddenly arise and swallow up the vessels which happen to be within the shock of these furious elements.

The Danish History relates, that in the year 1348, a great pestilence, which was called the *black plague*, depopulated a great part of the North. It carried off most of the sailors and merchants of Norway and Denmark who were engaged in the trade between Greenland and those kingdoms. About this period the navigation to Greenland became less frequent, and the traffic began to be discontinued. But the learned Wormius assured Peyrere, that he had read in a Danish manuscript, that down to the year 1484 there was a company of more than forty sailors, at Bergen, in Norway, who went every year to Greenland and brought back some valuable products. Some German merchants had come to Bergen for the purpose of purchasing these products, which the Greenlandmen were not willing to dispose of; and it is added, that the Germans, resenting this disappointment, invited the Greenland traders to a supper, at which they put them treacherously to death. But, as Peyrere remarks, this account has not much appearance of truth; nor is it probable, that the navigation between Greenland and

Norway was, at this period, so open as the above details would induce us to suppose. Those details are, besides, refuted by the following facts.

The revenue accruing from the province of Greenland was, in ancient times, appropriated to the domestic expenses of the Norwegian king; and no one could go to Greenland without the royal permission, upon pain of death. In the year 1389, Henry, Bishop of Garde, in Greenland, embarked for Denmark, and was present at the meeting of the States of that kingdom, which were held at Funen in the reign of Queen Margaret, who united the kingdoms of Denmark and of Norway under the same crown. At this time some Norwegian merchants, who had gone to Greenland without leave, were accused of having purloined the revenue which was reserved for the expenditure of the queen. The queen showed no lenity towards these merchants, and would have proceeded to take away their lives, if they had not made oath upon the Holy Evangelists that their voyage to Greenland was unpremeditated, and that they were forced to that destination by the violence of a sudden storm. They alleged that they had brought back only commodities which they had purchased, and that they had not in the least interfered with the revenue belonging to the queen. They were accordingly set at liberty; but the danger which they had escaped, and the more rigorous prohibitions which were issued, prevented any other individuals from that time from attempting to carry on any traffic with the interdicted coast.

Sometime after this the queen herself dispatched some vessels to Greenland; but of which no tidings were ever received; and they must consequently have perished. This disastrous expedition contributed to put an end to the intercourse with Greenland; and the queen having her attention occupied by her hostilities with Sweden, lost sight of this remote colony, or left it to its fate.

The Danish Chronicle relates, that in the year 1406, Eskild, Archbishop of Drontheim, wishing to exercise the same ecclesiastical authority over Greenland, which his predecessors had done, sent a prelate named Andrew, in order to succeed Henry, in the see of Garde, if he were dead, or to convey some intelligence concerning him if he were living. Nothing more was ever heard of Bishop Andrew, after his embarkation for

Greenland; nor were any farther tidings ever received of Henry, Bishop of Garde. After this, the intercourse between Norway or Denmark and Greenland, suffered an interruption from that period to the present; nor is there much probability that it will ever be renewed.

Queen Margaret was succeeded, upon the throne of Denmark, by Erick, of Pomerania, who gave himself little trouble about a settlement so remote as that of Greenland. His successor, Christopher of Bavaria, was employed during his whole reign in making war upon the Pomeranians.

The house of Oldenburg began its reign in Denmark in the year 1448. Christian, who was the first sovereign of that race, and of that name, neglected his dominions in the North in order to turn his attention to the South. He made a pilgrimage to Rome, where he obtained from the Pope a grant of the country of Ditmarsh, and permission to establish an academy at Copenhagen.

Christian I was succeeded by Christian II, who, at the period of his coronation, bound himself by a solemn promise to make every possible exertion to restore the intercourse between Denmark and Greenland, and to recover that settlement. But this monarch, instead of recovering what his predecessors had lost, himself lost part of what they had possessed. His tyrannical barbarities caused him to be expelled from Sweden, which Queen Margaret had united with the Danish and Norwegian crown. From Sweden Christian II retired into Denmark; but the same conduct which had occasioned his expulsion by the Swedes, soon led to his deposition by the Danes. It is on this account that he is represented with a shivered sceptre amongst the Danish kings.

Eric Valkandor, who had been chancellor to Christian II, and was a Danish gentleman of great and generous sentiments, had been made Archbishop of Drontheim. After the disgrace of his master, he retired to his archiepiscopal see, where he exerted himself with great zeal and activity in order to renew the communication with Greenland, and to discover the fate of that ancient settlement. This learned prelate made it his business to read all the books in which it was mentioned, to examine all the merchants and mariners who had any knowledge of it; and he also caused a chart to be formed of the route which was supposed to have been observed. He was on the point of putting his projects in execution, when, being suspected of favouring the cause of the deposed monarch, he was deprived of his archbishoprick, and banished from the Norwegian territory. The benevolent scheme, which he had formed, was thus disconcerted; and the hopes, which had been excited, vanished in disappointment. The good Archbishop Valcandor retired to Rome, where he ended his days.

Frederick I was succeeded by Christian III, who had an expedition fitted out for the discovery of the lost settlement in Greenland; but this proved as abortive as similar attempts had previously been. This monarch now repealed the ordinances which his predecessors had established, by which all communication with Greenland had been strictly prohibited, without a special permission from the crown. The intercourse was now rendered free, without any limitations or restraints. But this act of royal grace came too late to be of any use; for the Norwegians at this period had degenerated from the enterprizing valour of their ancestors; and they were, at the same time, so impoverished that they did not possess the means of equipping any vessels for such a difficult and hazardous undertaking.

Frederick II entertained the same project as his father, Christian III, and he dispatched *Magnus Heigningsen* to attempt the discovery of Greenland. This Magnus Heigningsen, if the relation be not fabulous, actually discovered the long lost land, but was prevented by the operation of some mysterious cause from reaching the shore. His ship, without any visible cause, was stopped in its course, though in the midst of deep water and a fresh breeze, without any obstruction from the ice. As this Magnus Heigningsen could not advance any farther he was happy to be able to retreat; and he accordingly sailed back to Denmark. When he got back to that country he published an account of what had happened to his ship; and pretended that its farther progress had been stopped by a great loadstone at the bottom of the sea.

The Danish Chronicle, of which Peyrere has made such liberal use, gives the following account of the expedition of Sir Martin Frobisher to Greenland in 1576.

Frobisher set sail from England in the year just mentioned, and discovered the coast of New Greenland, but did not make any landing till he returned with another expedition in the following spring. The inhabitants of that part of the coast where he disembarked, abandoned their dwellings, and fled in different directions at the approach of the English. The alarm of some of these natives appears to have been so great that they clambered up to the tops of some rocky precipices, from which they threw themselves into the sea.

The English, who found it impossible to allay the suspicions, or conciliate the confidence of these savages, took possession of the huts which they had deserted. They were, in fact, tents formed of sealskins, stretched upon four poles, and sewed together with sinews instead of thread. All these tents had two entrances, one of which fronted the West, and the other the South; but they were closed against the winds from the East and the North, by which they were liable to be the most incommoded.

The English discovered in these cabins only an ancient matron, who appeared a picture of hideous deformity, and a young woman, who was in the family way, and had a little child holding her hand. These two last they carried off, regardless of the opposition of the old beldam, who set up a frightful howl. Departing from this point, they steered along the Eastern coast, where they beheld a marine monster as large as an ox, with a horn projecting from the snout of more than two yards in length, which they took for the unicorn. Proceeding in a North-east direction, they landed on another part of the coast of Greenland, which they discovered to be subject to earthquakes, that threw great rocks down into the plain. Here they found some gravel abounding, as they imagined, with particles of gold, of which they carried off a considerable quantity.

They spared no pains to conciliate the natives of this part of the coast, who themselves made a show of a desire to maintain an amicable correspondence. But these demonstrations of friendship appear to have been designed only to put the English off their guard; for, when Frobisher had landed, he was suddenly attacked by a body of savages, who had concealed themselves behind a bank for that purpose. He retreated to the shore and eluded their machinations. The savages, however, still imagined that the strangers might be caught in the snare; and in order to entrap them, they scattered pieces of raw flesh along the shore, as they would have done to allure dogs. Finding this attempt fail, they had recourse to another stratagem. They carried a lame man, or at least one who feigned to be lame, down to the sea-side; and, having left him there, they went away and kept themselves entirely out of sight. They supposed that the English would make an attempt to carry off this lame man in order to serve them as an interpreter, or to procure some intelligence by his means. But Frobisher, who suspected some deception, ordered a shot to be fired over his head, when he instantly sprung up upon his legs and ran away with precipitate velocity.

The savages now appeared in great numbers, and assailed the English with a shower of arrows and stones; but they were soon repulsed by a discharge of great and small guns.

The native Greenlanders are represented as perfidious and cruel, neither to be softened by caresses nor moved by benefits. This, however, is the character of very imperfect knowledge and limited observation. They are described as plump in their appearance, active in their limbs, and with an aspect of olive hue. Some of them are reported to be as black as negroes. Their clothes are made out of the skin of the seal, and sewed with sinews. The women wear their hair loose, but throw it back behind their ears in order to show the face, which they paint blue and yellow. They wear no petticoats, but short trousers made of fish-skin, drawn one over the other; in the pockets of which they carry their knives, little mirrors, and the working materials, which they procure from foreigners or obtain from the wrecks which may happen upon their coasts. The shirts or chemises of both sexes are made from the intestines of fish, and sewed with fine sinews. They wear their clothes loose, and gird them with a belt made of fish-skin. They are disgustingly dirty, and covered with vermin. Their criterion of wealth is the number of bows and arrows, of slings, boats, and oars, which an individual may possess. Their bows are small, their arrows thin and armed at the end with a sharp point of bone or horn. They are expert in the use of the bow and the sling; and in killing fish with the spear. Their little boats are covered with sealskin, and can hold only one man. But they have larger boats formed of wood, covered with the skin of the whale, and which will carry twenty men. Their sails are made of the same materials as their shirts; or of the intestines of fish fastened together by fine sinews. And though they make use of no iron in the construction of their canoes or boats, they are put together with so much skill, and so well compacted, that in them they venture out into the wide ocean with perfect security. They have no venomous reptiles or insects; but are sometimes infested with swarms of gnats. They make use of very large dogs for the purpose of drawing their sledges. All the fresh water which they possess they procure from the melted snow.

Such are the principal particulars which are detailed in the Danish account of Frobisher's voyage. We will now proceed to relate some attempts of the Danes to renew their intercourse with Greenland, subsequent to those which have been previously mentioned, and which proved abortive.

Christian IV resolved, if possible, to signalize his reign by the discovery of that lost settlement, which his father and grandfather had sought in vain. For this purpose he sent for an experienced mariner from England, who had the reputation of being well acquainted with the Northern Ocean, and with the route to Greenland. Having procured this skillful auxiliary, whose name was John Knight, the Danish monarch equipped three stout ships, which he put under the orders of Godske Lindenau, who sailed from the Sound on the breaking up of the ice in the year 1605. The Englishman, who was appointed to the command of one ship, having reached the latitude he wished, steered his course to the South West in order to avoid the ice and to make the land with less risk. The Danish admiral Lindenau, thinking that the English captain was deviating from the right track by keeping to the South West, continued his route to the North East, and arrived on the coast of Greenland without either of the other ships. Admiral Lindenau had no sooner come to an anchor, than a number of savages put off their boats from the shore to visit his ship. The admiral gave them a very hospitable reception, and made them a present of some wine, which, however, was not agreeable to their taste; and they manifested signs of their dislike. They saw some whale oil, which they expressed a desire to have; and it was accordingly poured out for them in large mugs, which they drank with avidity and delight.

These savages possessed a number of skins of the fox, the bear, and the seal, with many horns in pieces, ends, and trunks, which they exchanged with the Danes for knives, needles, looking-glasses, and trifles of different kinds. They showed no desire for gold or silver money, the offer of which provoked their ridicule or excited their contempt. They manifested on the other hand a passionate eagerness for every article of steel manufacture, which they were willing to purchase by the sacrifice of their greatest valuables, as of their bows and arrows, their boats and oars. When they had nothing more to offer in exchange, they stripped themselves to the skin, and offered to make away with all the clothes they possessed.

Godske Lindenau remained three days in the road, but it is not said that he once went ashore. He was probably afraid of trusting the lives of the small number of persons he had with him in the midst of such a mass of savages, by whom they were so greatly outnumbered.

He took his departure upon the fourth day; but before he set sail he secured two of the natives on board his ship in order to carry them to Denmark; but they made so many violent efforts to escape, that it became necessary to secure them by cords in order to prevent them from plunging into the sea. When the savages upon the beach saw two of their countrymen made prisoners and fastened to the deck of the Danish vessel, they discharged a shower of stones and arrows upon the Danes, who were obliged to terrify them to a distance by firing off one of their great guns. The admiral returned to Denmark by himself, without knowing what had befallen the other two ships, with which he had originally embarked.

The Danish account of this expedition says, that the English captain with the two Danish vessels, which had separated from that under Lindenau, reached the coast at the Southern extremity of Greenland, or Cape Farewell. It is also certain that the English commander entered Davis's Straits, and coasted along the shore to the East. He discovered a number of good harbours, a fine country, and verdant plains. The savages in this part of Greenland carried on some traffic with him; as those upon the other side had done with Lindenau; but they exhibited more distrust; for they had no sooner received the Danish commodities in exchange for their own than they took to their boats with as much precipitation as if they were pursued by an enemy.

The Danes armed themselves for the purpose of making a landing in one of the bays. The soil, where they went ashore, appeared to be a mixture of sand and rock, like that of Norway. Some fumes exhaled from the earth made them suppose that there were mines of sulphur in the neighbourhood; and they found many pieces of silver ore, which yielded twenty-six ounces of silver to the hundred weight of ore.

The English captain, who discovered many fine harbours or bays along this coast, gave them Danish names, and before his departure made a chart of what he had seen. He also directed four of the best formed savages, whom the Danes could seize, to be conveyed on board his ship. One of these four natives became so outrageous, that the Danes, not being able to haul him along, knocked him on the head with the but end of their muskets. This intimidated the three others, who followed without farther resistance.

But the natives of the place, who had beheld one of their companions put to death, and three made prisoners, united themselves in a body to avenge the one and to rescue the others. They pursued the Danes to the shore in order to execute these resolutions, and to prevent their embarkation. The Danes, however, saved themselves and their boats by a timely use of their fire-arms, which diffused great terror among the enemy. They now made good their retreat to their ships, and returned to Denmark with the three captured Greenlanders, whom they presented to the king, and who were found to be much better made and more civilized than those whom Godske Lindenau had imported. They also differed in manners, language, and dress.

The Danish monarch, who was gratified by the result of this first expedition, dispatched the same Admiral Lindenau to Greenland with five stout vessels in the following year, 1606. He departed from the Sound upon the 8th of May; having on board his ship the three savages whom the English captain had conveyed away, in order to serve as interpreters and guides. One of these savages fell sick and died during the voyage; and his body was thrown overboard. Godske Lindenau took the same route which the English captain had observed, and passed by Cape Farewell into

Davis's Straits. One of his five ships was lost sight of in a fog; but the four others arrived in Greenland. The natives showed themselves in great numbers upon the coast, but manifested no inclination to trade, or to trust the Danes, who, in their turn, showed the same want of confidence. This obliged the latter to proceed higher up the coast, where they discovered a finer harbour than that which they had left; but they found the natives as suspicious and intractable as at the former station, and indicating a determination to resort to force if the Danes attempted to land.

The Danes, not willing to hazard a landing in such inauspicious circumstances, sailed to a greater distance. As they proceeded along the coast, they met some of the natives in their canoes. They surprised six of these at different times, and took them on board along with their canoes and little equipments.

The Danes, having afterwards cast anchor in a third bay, one of the attendants of Godske Lindenau, who was a hardy and enterprising veteran, solicited the permission of his master to proceed alone to the shore, in order to reconnoitre the land, and, if possible, to establish some intercourse with the savages. But this unfortunate valet had no sooner set his foot upon the beach than he was seized, stabbed, and hacked in pieces by the natives; who, after this atrocity, retired out of the reach of the Danish guns.

These savages had knives and swords made of the horns or teeth of that fish which they call unicorn, and which they ground to an edge upon a stone; nor were they less sharp than if they had been made of iron or steel.

Godske Lindenau, not finding it practicable to establish any amicable communication with the people of this district, set sail for Denmark; but of the six Greenlanders whom he had recently forced on board, one was pierced with such regret at the thought of never more seeing his native home, that he threw himself into the ocean in a paroxysm of despair. Upon their return the Danes had the pleasure of rejoining the fifth ship, which had disappeared in a fog; but they had been only five days together when they were all separated by a storm; and a month elapsed before they could re-unite when the tempest had passed away. They arrived at Copenhagen upon the 5th of the following October, after having experienced many awful perils and hairbreadth escapes.

The King of Denmark, who deserves praise for his perseverance, now determined upon a third expedition to Greenland. He accordingly ordered two large ships to be fitted out, which he placed under the command of a Captain Karsten Richkardisen, a native of Holstein, whom he furnished with some sailors from Norway and Iceland that were acquainted with the navigation. These vessels sailed from the Sound on the 12th of May, but the Danish Chronicle has not stated in what year; nor was it known to Peyrere. On the 8th of June Richkardisen discovered the high points of the Greenland mountains; but he was prevented from landing by the rocks of ice which ran out far into the sea and rendered the coast inaccessible. He was therefore obliged to return without accomplishing the object of his voyage, as he despaired of being able to penetrate the icy barrier which blockaded the shore. No similar attempt has hitherto been successful; and the Eastern coast of Greenland, though for several centuries well known to, and habitually visited by, the Norwegians and Danes, is, at present, a *terra incognita*, notwithstanding the spirit of European adventure and the zeal of modern discovery.

The King of Denmark caused particular attention to be paid to the three savages who had survived the preceding, and the five who had been imported by the last expedition to Greenland. They were fed upon milk, butter, and cheese, as well as upon raw flesh and raw fish, to which they had been accustomed at home. They appeared to have an invincible repugnance to our baked bread and dressed meat; nor did they relish any kind of wine so much as the oil and grease of the whale. They often turned a wishful and desponding look to the North; and sighed so anxiously to return to the place of their nativity, that, whenever they were watched with less vigilance than usual, those who had an opportunity seized any boat that was at hand and put to sea, regardless of the dangers they had to encounter. A storm once overtook some of these intrepid adventurers at ten or twelve leagues from the Sound, and forced them back to the coast of Schonen, where they were made prisoners by the peasantry and conveyed back to Copenhagen. This caused them to be guarded with more rigour, and kept under greater restraint. But three of them fell sick and died of grief.

Five of these savages were alive and well when a Spanish Ambassador made his appearance in Denmark; and the Danish Monarch, in order to divert this stranger, caused these native Greenlanders to exhibit their manœuvres in their little canoes upon the sea. The Spanish Ambassador was quite delighted with the address which they displayed, and with the extraordinary celerity with which they glided over the waves. He made a present in money to each of the savages, which they expended in equipping themselves in the Danish fashion. They were accordingly seen booted and spurred, with large feathers in their hats; and in these habiliments they proposed to serve in the cavalry of the Danish King.

But these high spirits of the Greenlanders lasted only for a short time; for they soon relapsed into their usual melancholy. They became entirely absorbed with the idea of returning to their native country; and two of them having obtained possession of their little boats put out to sea. They were pursued, but only one of them was taken, and the other probably perished in the waves; for it cannot be supposed that he ever returned to the land of his fathers. With respect to one of the savages, it was remarked, that he shed tears whenever he beheld a child at the breast; from which it was supposed, that he had left a wife and children at home.

Of these surviving savages two pined away with regret. The two others lived ten or twelve years in Denmark after the decease of their companions. No pains were spared to reconcile them to their condition, but without success. One of them died of an illness, brought on by being employed in diving for the pearl muscle, during the depth of winter. His companion, who was inconsolable for his loss, again seized a boat and made an effort to escape from captivity. He had passed the Sound before he could be retaken, but he lived only a short time after this last attempt to recover his liberty.

Peyrere says, that an attempt was made to convert these savages to the Christian faith, but that they could never be brought to learn the Danish language; and he remarks, with much simplicity, that "la foi estant de l'oüye, il fut impossible de leur faire comprendre nos mysteres." "Faith," says he, "coming from hearing, it was impossible to make them comprehend our mysteries." He adds, that those who narrowly watched their actions often saw them lift up their eyes to Heaven, and worship the Sun.

The Danish Monarch desisted from any farther attempt to discover Old Greenland; but some merchants at Copenhagen formed themselves into a Greenland Company, for the purpose of establishing a traffic with that part of the world. In 1636 this Company fitted out two ships, which visited that part of the coast of New Greenland which is washed by Davis's Straits. When they cast anchor, eight savages came off to them in their little canoes. The Danes had displayed their knives, mirrors, and other articles upon the deck, to which the savages had also conveyed their furs, skins, and fish horns; but a gun having been inconsiderately fired, in order to celebrate the drinking of some particular health, these native traders were so frightened that they instantly leaped into the sea, from which they did not emerge till they had proceeded to two or three hundred yards from the ship.

The Danes at last succeeded in appeasing the apprehensions of the Greenlanders, and in alluring them again on board their vessels. The Danish commander having remarked an inlet of the coast where there was a bank of sand, which bore a strong resemblance to gold, his cupidity made him imagine, that he had discovered a mine of wealth. He lost no time in filling his ship with this fancied gold dust, and made the best of his way to Denmark, exulting in dreams of visionary opulence.

But the master of the Greenland Company, who was less credulous than the captain of the expedition, having caused this precious sand to be examined by the goldsmiths at Copenhagen, they were not able to extract from the whole mass a single particle of gold. The captain was accordingly ordered, to his great mortification, to throw the whole of this valuable lading into the sea.

In this last expedition to Greenland the Danes secured and carried off two of the natives before they left the coast. When they had reached the open sea, the Danes released these captives from their bonds, when, finding themselves free from restraint, the love of liberty prevailed over every other sentiment, and they plunged into the waves in order to regain their native shore. But that shore was too remote for them to reach, and they perished in the vain attempt. It is pleasant to contemplate that sentiment, which attaches us to our native land, operating alike in all regions and climes, and attaching the human being to a country of almost invincible sterility and perpetual frost, as well as to one where there are the richest products and the most genial seasons.

In the year 1654 a ship was sent to Greenland, under the command of David Nelles, the success of which terminated in carrying off three native women from the open part of the Eastern coast. The last voyage, which was not more successful than the preceding, was made in the year 1670. This expedition was fitted out by order of Christian V, and was commanded by Captain Otto Axelson; but Crantz[19] says, "We have no account of its issue;" and, according to Torfæus, Axelson never returned to tell what he had seen.

None of the expeditions which have sailed from Denmark, or other countries, have been successful in recovering the knowledge of that part of the Eastern coast which was peopled by settlers from Iceland and Norway, and is denominated Old Greenland. In the account which the Icelandic Chronicle gives of the ancient route, it is stated, that half way between Iceland and Greenland there was a cluster of little islands, or rocks, called Gondebiurne Skeer, which were inhabited by bears. The drifting ice has probably collected round these islands, and been so petrified by successive accumulations as to become impenetrable to the sun.

Peyrere, whose account of Greenland has been generally followed in this Introduction, tells us, that he was once inclined to believe, that Godske Lindenau had actually reached the coast of Old Greenland in his first voyage, and that the savages whom he carried off were descendants of the first Norwegian settlers, whose remains have been so anxiously sought. But this impression was effaced by the information of many persons who had seen these savages at Copenhagen, and who assured him, that they had not the smallest resemblance to the Danes or Norwegians in their language and manners, and that the Danes and Norwegians could not understand a word that these native Greenlanders uttered.

---

[19] Vol. i. p. 278.

In the expedition to Greenland, which was undertaken in the year 1636, the natives upon the western coast, who had some traffic with the Danes, were asked, whether there were any inhabitants like themselves beyond the mountains which were seen in the distance. The savages replied by signs, that there were more people beyond the mountains than there were hairs on their heads; and that they were men of large stature, with great bows and arrows, who destroyed everybody that came in their way.

The knowledge which the Danes have at any period acquired respecting the people or the products of Greenland, never extended beyond a narrow slip of territory along the coast. They knew nothing of the remote interior of the country from actual observation; and their settlements occupied only a very small comparative portion of the whole. Much is still left to be explored; but the nature of the country itself opposes such an accumulation of obstacles to the research of the traveller, that they are not soon likely to be overcome. More, however, of the coast will probably soon be discovered than has ever previously been explored; or, if explored, it has at least been concealed for many centuries. When the enterprizing spirit of an English navigator is directed to that quarter of the world, we feel a firm confidence, that nothing will be left untried, which skill or courage can effect, to extend our acquaintance with these Northern regions, and to make valuable additions to our present stock of information respecting the countries in the more immediate vicinity of the North Pole.

SKETCH OF THE LIFE OF HANS EGEDE

THE Author of the present Volume was born in Denmark, on the 31st of January, in the year 1686. He was educated for the Christian ministry, and became pastor to a congregation at Vogen, in Norway, and appears for some time to have exercised the same functions at Drontheim, in that kingdom. In an early period of his ministry he was seized with a strong desire of making himself acquainted with the fate of the Norwegian families who had formerly been settled in Greenland, and of whom no intelligence had been received for several centuries. All the inquiries which he could make led to the conclusion, that that part of the coast where these settlements had formerly existed had been rendered inaccessible by the ice; and that the ancient settlers had been destroyed either by the effects of the climate or the hostility of the natives. But these unfavourable representations did not repress the ardour of Egede to embark in this perilous undertaking; and either to discover the old Norwegian settlements, or to form a new one, and to devote his life to the instruction of the barbarous and uncivilized Greenlanders in the salutary truths of the Christian doctrine.

He was a man of warm temperament, and mingled with such a portion of enthusiasm as does not readily suffer its exertions to be relaxed by difficulties, or the hopes which it has conceived to be extinguished by inauspicious circumstances. For many years he attempted in vain to interest the Danish government in the furtherance of the scheme which he had conceived. His memorials were disregarded, and his proposals were considered as visionary and impracticable. But at last Frederick IV, King of Denmark, issued an order to the magistrates at Bergen to make inquiries of all the masters of vessels and traders, who had been in Davis's Straits, concerning the state of the traffic with Greenland; and, at the same time, to learn their opinion about forming a new settlement upon that coast. But the answer which they returned was not at all favourable to the wishes of our author, and the project seemed never likely to be accomplished.

After more ineffectual attempts, his perseverance at last triumphed over every obstacle; and he persuaded some merchants and others to subscribe some small sums, out of which he collected a capital of about 2000*l*. Of this inconsiderable sum he himself had furnished about 60*l*, which constituted his little all. With these slender means, which seemed totally inadequate to the undertaking, a ship was purchased, called the Hope, in which Egede was to be conveyed to Greenland, and to lay the foundation of the meditated establishment. But, in the spring of 1721, the Danish monarch, who had been brought to think more favourably of the expedition, appointed Mr. Egede to be pastor of the new colony, and missionary to the Heathen, with a pension of 60*l* a year, and 40*l* for his immediate exigencies.

Egede embarked for Greenland, with his wife and four small children, upon the 12th of May, 1721; and he landed in Ball's River, in the 64th degree of North latitude, upon the 3d of July, in the same year. The company on board the ship consisted of forty persons. They lost no time in building a house of stone and earth, upon an island near Kangek, which they called Haabets Oe, or Hope Island, after the name of the ship in which they had made the voyage.

The conduct of Egede as a missionary deserves the highest praise. He conciliated the confidence of the natives, ministered to their wants, learned their language, and gradually introduced some additional rays of intellectual light into their minds.

"As soon," says Crantz, vol. i. p. 286, "as he knew the word *kina*, i. e. what is this? he asked the name of everything that presented itself to the senses, and wrote it down." But his children, by continually conversing with the children of the natives, learned the language, particularly the pronunciation, with much more facility than himself; and he was enabled to make considerable use of their proficiency in the vernacular tongue of the country, in promoting the purposes of his mission.

Upon the death of Frederick IV, and the accession of Christian VI, the Danish government, dissatisfied with the expense which the settlement in Greenland had occasioned, and the faint prospect which appeared of any adequate remuneration from the trade with that country, issued, in 1731, a mandate for the relinquishment of the colony, and the return of the settlers. But this zealous missionary resolved not to abandon the good work which he had begun; and though most of the settlers left the coast in the ship which had been sent to conduct them home, he remained behind with ten seamen whom he had persuaded to adopt the same determination. The Danish monarch, either sympathising with his constancy, or moved by his entreaties, assisted him with some supplies in the following year; and in the year 1733 he was cheered by the assurance that the mission should be more effectually supported, and the trade with Greenland more vigorously prosecuted than it had hitherto been.

When the advanced age, or rather the growing infirmities of Egede, which had been increased by the corporeal toils he had undergone, and the mental solicitudes he had experienced, no longer permitted him to continue his former occupation, his eldest son Paul became his successor in the mission. After an abode of fifteen years in this sterile region and inclement climate, he returned to Copenhagen in the year 1736. Though he had relinquished the mission, he was not inattentive to its interests; for he devoted much of his time, after his return, to the instruction of young missionaries in the language of Greenland. He also composed a grammar and a dictionary of that language, into which he translated the New Testament for the use of the mission and the benefit of the natives. He published the Description of Greenland, which is contained in the present

Volume, at Copenhagen, in the Danish language, the year preceding his death, which took place in 1758.[20]

---

[20] The Moravian mission in Greenland began in the year 1733. The brethren of this mission have two settlements or villages upon the Western coast. One of these, which is called New Herrnhut, is on Ball's River; and the other, which is denominated Lichtenfels, is at the distance of thirty-six leagues from the first, and more to the South. Crantz says, that, at his departure from Greenland, four hundred and seventy Greenlanders were living at New Herrnhut in sixteen houses. The brethren themselves describe this place as a sort of green Oasis in a cheerless desert. "No one," says one of the missionaries, "would expect to find such a pleasant place in such an unpleasant land. The country consists entirely of bald rocks, thinly interspersed with spots and veins of earth, or rather sand. But our house, area, garden, &c. look very regular and decent, and all the adjacent land round about the place, where once not a blade of grass grew in the sand, is now enrobed with the most beautiful foliage, so that New-Herrnhut may be called a garden of the Lord in a most frightful wilderness." Crantz, vol. i, p. 162, 163. In p. 399, of the same volume, Crantz extols the soft beauty of this little Greenland village, compared with the rugged sterility around. "On the spot," says he, "that formerly consisted of nothing but sand, nay, on the very rocks, grows now the finest grass, the ground being manured for so many years with the blood and fat of their seals. And when the Greenlanders live in their winter houses, one may see every evening, yea, throughout the whole night, a beautiful illumination, which is the more agreeable as the houses stand in two parallel lines, are of equal height, and have light in all the windows."

PREFACE[*]

A friend of mine, who lived some time in Greenland, published (unknown to me) some years ago, a Description of Greenland, under the title of A New Survey of Old Greenland, which, not long after my arrival in those parts, I had sketched, to satisfy some of my friends, according to

---

[*] This is an edited, reformatted and augmented version of *A Description of Greenland* by Hans Egede, originally published by T & J Allman, dated 1818. The views, opinions and nomenclature expressed in this book are those of the author, and do not reflect the views of Nova Science Publishers Inc.

the knowledge I then had acquired; but having since that time got a fuller light in these matters, partly by my own observations and partly by those of my Son Paul Egede, who has been four years missionary in the North West colony of Greenland, I have found it necessary to perfect and enlarge this little Work in embryo, under the same title that it made its first appearance, with some useful Additions, and with a new contrived Map of the country, that the reader may the better comprehend what he finds in this Sketch.

Though Greenland be a country of a vast extent, yet it affords but a narrow field for any observation or remarks of consequence; there being no strong or well built towns to meet with; no well ordered polity or civil government; no fine arts and sciences, or the like; but only a number of mean, wretched, and ignorant Gentiles, who live and improve the land according to their low capacity.

I must own, that Greenland, in its present state and condition, compared with other countries, is but very mean and poor, though not yet so despicable and wretched but it may, using care and industry, not only richly maintain its own inhabitants, but also communicate to others out of the remainder of its products.

As for the land in itself, it yields little or nothing, not being manured or cultivated, but lies altogether waste and untilled; nevertheless the ancient histories and accounts, yet extant, of the land, make it appear, that it is not unfit for several products; and therefore I do not question but it might retrieve the loss of its former plenty and fruitfulness, should it come to be well settled again, and cultivated. But as to the seas, they yield more plenty and wealth of all sorts of animals and fishes than in most other parts of the world, which may turn to very great profit; witness the exceeding great riches many nations have gathered, and are still gathering, from the whale fishery, and the capture of seals and morses, or sea horses.

Thus it is confessed, that Greenland is a country not unworthy of keeping and improving. And this has been the well grounded opinion of our late monarchs of Denmark, and many of their chief counsellors, who have made so much of Greenland, that they have spared no costs in fitting out several ships for its discovery, of which hereafter farther notice shall be taken. This discovery has been chiefly undertaken to the end, that the

Christian religion, which has been unfortunately worn out in these parts of the world, might again be re-established, and the poor inhabitants, viz. the offspring of the old Northern Christians, if through God's mercy any such may yet be found there, as true subjects to Denmark and Norway, might be assisted and comforted both as to body and soul. And although these most laudable endeavours of those glorious monarchs, of pious and blessed memory, have not had all the success one could desire, yet they have opened the way for fresh attempts of the same nature, which (God be thanked) have not been lost, inasmuch as the Western coast of Greenland (by the Danes called Westerbygd) not only has been fully discovered, but also several new lodges have been there erected, and the holy word of God has been preached, with God's blessing, to these ignorant Heathens, that dwell in those places where Christianity has been quite extinct and forgot. All this ought to encourage us to continue our endeavours to discover the Eastern shore, where it is confessed the chief colony has been seated; and perhaps the offspring of the old Norwegians and Icelanders may be recovered; which I do not think impossible, provided we go on in the right way, as I hope to show in the following treatise.

How praiseworthy and glorious an enterprize would it be, to undertake so great and wholesome a work, chiefly in regard to these unhappy people, who, by a just judgment of God, now for upwards of three hundred years, have been debarred all communication with Christians; which to remedy not only our civil but Christian duty obliges us. It becomes us therefore heartily to pray God Almighty, that he will be pleased to appease his wrath kindled against these poor wretches, and to disclose to our most gracious sovereign, and to other well intentioned Christians, the best way and means to this country's discovery and happy restitution. And though we should fail of success, in still meeting with the aforesaid offspring of the old Norwegian and Iceland Christians, who, for aught we know, may be all extinct and destroyed, as we found it on the West coast; yet, for all that, I should not think all our labour lost, nor our costs made to no purpose, as long as it may be for the good and advantage of those ignorant Heathens, that live there; to whom we have reason to hope our most gracious sovereign will also extend his fatherly clemency, and Christian zeal, to provide for their eternal happiness, as he so graciously has done for those on the Western shore; seeing that by these means the old ruined places might anew be provided with colonies and inhabitants, which would prove no small advantage to the king and his dominions. This my well-meant project, that God in his mercy will advance and promote, to the honour of his most Holy Name, and the enlightening and saving of these poor souls, is the sincere desire of

Hans Egede

Chapter 1

OF THE SITUATION AND EXTENT
OF GREENLAND

Greenland lies but forty miles to the West of Iceland, beginning from 59° 50′ North Latitude. The Eastern coast extends itself in the North as far as Spitzbergen, between 78° and 80°; which is thought to be an island, separated from the continent of Greenland. The Western shore is discovered as far as seventy odd degrees. Whether it be a large island, or borders upon countries to the North, is not yet found out; there seems great reason to believe it is contiguous to America on the North West side; because there we meet with the bay or inlet, which in the sea charts is called Davis's Straits, from an Englishman, who in the year 1585 was the first discoverer of it; and is yearly frequented by ships of different nations, on account of the Whale Fishery: but nobody as yet has been able to find out the bottom of it. And according to the notice we have endeavoured to gather from those Greenlanders who live farthest to the North, there is either but a very narrow passage between America and Greenland, or, as is most likely, they are quite contiguous:[21] and I am the more inclined to believe this, because the farther you go Northward in the said Strait, the lower is the land; contrary to what we observe where it borders on the seas or main ocean, it never wants lofty promontories. It has been the commonly received opinion, of a long standing, that Greenland borders upon the Asiatic Tartary and Muscovia on the North East: what confirms them in this notion is an old story they give credit to, that a certain Harrald goat did travel by land, over mountains and rocks, from Greenland to Norway, bringing along with him a she goat, of whose milk he lived on the journey; by which he got the surname of Harrald goat. Furthermore, the ancient Greenland Christians, in their Chronicles, relate, that there were come to them from the Northern parts, foreign rein deer and sheep, marked upon the ears, and with some marks tied to their horns; from which they concluded, that the Northern parts of Greenland were also inhabited.—Vid. Theodore Torlaccius. But the contrary is proved by later experiments made by the navigation of Dutchmen and others to the North.—See Zordrager's Greenland Fishery, Part ii, ch. 10.

Greenland is a high and rocky country, always covered with ice and snow (except on the sea side, and in the bays or inlets) which never thaws nor melts away. You may judge of the height by the prospect they yield at more than twenty Norway miles distance from the shore. The whole coast

---

[21] According to the relation and opinion of those Greenlanders, that inhabit the gulf of Disco, in 69°, Greenland is an island, which they infer from the strong current that runs from the North, and keeps the ice open even into the midst of the sea: they will also tell you, they have spoken with people different from themselves on the other side of the ice, and hailed them. Their language, they say, is the same, but the persons different, so that a small strait only divides Greenland from America. The said straits are so narrow, that men on both sides can shoot at once one and the same fish. The continent farthest to the North is all covered with ice; the islands only uncovered, where rein deer, and also geese and other wild birds, are found in great numbers.

is surrounded with a vast number of large and small islands. There are a great many inlets and large rivers to be met with, among which the principal is called Baal's River, in 64°, and has been navigated eighteen or twenty Norway miles up the country; where the first Danish lodge was settled in the year 1721. In all sea charts you will find laid down Frobisher's Strait and Baer Sound which they pretend, form two large islands, adjacent to the main land; which I think are not to be found, at least not upon the coast of Greenland; for I could not meet with anything like it in the voyage I undertook in the year 1723 Southward, going upon discoveries; though I went as far as to 60° that way: but at present the newer charts lay them down, the Northern strait in 63°, and the Southern in 62°. Some of the ancients, whom Thormoder follows in his Greenland History, place them between 61° and 60°. So that the charts differ mightily in this particular. Besides this, there is not a word or a syllable mentioned in our ancient records of Greenland of the aforesaid two straits and large islands: they only inform us, that after the old Norwegians and Icelanders had begun to settle colonies on the East side of Greenland, over against Iceland, they continued to spread themselves all along the shore and in the bays, as far as Baal's River, where they stopped, and where we find many ruins of the old Norwegian edifices. And whereas I myself have lately met with so many stone buildings, so far to the South, I think my conclusion is good, that the land upon which these houses stand is no particular island, but contiguous to the main. It is therefore very reasonable to believe, that whereas the ancients took notice of, and so accurately described, all those bays and islands that were inhabited, they would not have passed by in silence these two large islands upon which such stately buildings were erected. And for this reason I have hereto joined a new map or delineation of Greenland, to show the contiguity of the East and West Greenland, agreeably to other new charts of Thormoder and others, which I follow, as far as I find them not contradictory to the description of the ancients and to my own experience.

Chapter 2

FIRST SETTLEMENT OF GREENLAND, WITH SOME THOUGHTS ON THE EXTINCTION OF THE NORWEGIAN COLONIES; AND WHETHER ON THE EAST SIDE NO REMAINDERS MAY BE FOUND OF THE OLD NORWEGIANS: ALSO, WHETHER THE SAME TRACT OF LAND CANNOT BE RECOVERED

It is undoubted that the ancients, not so much driven by any necessity or compulsion as led by a natural and inbred curiosity, embarked upon many strange ventures; as for instance, to discover and settle colonies in so many formerly quite unknown and uninhabited countries, to whose discovery what particular accidents have most contributed we learn by the several histories and descriptions thereof. For the Almighty and good God, who has not in vain created the vast globe of the Earth, has also not intended, that any part or province of it should lie buried in eternal oblivion, useless to mankind. And that Greenland by such means has been discovered and inhabited by our old Norwegians and Icelanders, we are fully informed by the annals of Iceland; where we read, that the brave and valiant Erick Raude (or red) who was the first discoverer of this country, after he, in company with several other Icelanders, in the year of our Lord 982, by mere casualty fell in with the land, and had taken a survey of its present state, he returned to Iceland the next year, 983, spoke much in commendation of the land, calling it the Greenland, by which he persuaded many of his countrymen to follow him thither, in order to find out places fit for dwelling, and to settle there.[22] They no sooner were arrived and settled here, but they found God was come along with them; I mean the saving knowledge of his most holy Word. For the said Erick Raude's son, called Leif, after he had been instructed in the Gospel truths by King Olaf (who was the first Christian king of Norway), brought along with him from Norway to Greenland a priest, who taught and christened all the inhabitants of the country. Thus this country has first been settled by Norway and Iceland colonies, which, in after-times, have increased and been provided with many churches and convents, bishops and teachers; which lasted as long as the correspondence and navigation continued between them and Norway, until the year 1406, when the last bishop was sent over to Greenland. Yet the Norwegians were not the original natives of the land; for, not long after their arrival, they met with the old inhabitants, a savage people dwelling on the Western shore, originally descended from the Americans, as may with great probability be gathered from the agreement of their persons, customs, and habits with those who dwell to the North of Hudson's Bay; as likewise while those, that inhabited the Northern parts (now known by the name of Davis's Straits), advanced nearer and nearer to the South, and often made war upon the Norwegians. Concerning the cause of the ruin and total destruction of that so well established Norwegian colony there is nothing found upon record; the reason of which I think to

---

[22] Historians disagree about the time of the first settlement of Greenland. The Icelanders (as we have mentioned) will have it to be in the year 982-3. But Pontanus, in his Danish History, refers it to the year 770; making his assertion good by a bull of Pope Gregory the IVth, who in the year 835 sent to Bishop Ansgarius, wherein the propagation of the Gospel is recommended to him, as archbishop of the Northern Countries, and especially of Iceland and Greenland.

be, that after all correspondence and navigation ceased between Greenland and Norway, partly by the change and translation of the government in Queen Margaret's reign, and partly by the next following continual wars between the Danes and Swedes, which caused the navigation to those parts to be laid aside, and chiefly by the great difficulty and innumerable dangers of such navigation; which several causes cut off all intelligence, that might be had of that country's state, as may be seen in Pontanus and Claudius Lyscander.

The ancient historians divide Greenland into two parts or districts, called West Bygd, and East Bygd. As to the West district, which is said to have contained four parishes, and one hundred villages, all we find in the ancient histories amounts to this, viz. that in the fourteenth century it was sorely infested by a wild nation called Schrellings, and laid so waste, that when the inhabitants of the Eastern district came to the assistance of the Christians, and to expel the barbarous nation of the Schrellings, who were fallen upon the Christians, they found to their great astonishment the province quite emptied of its inhabitants, and nothing remaining but some cattle and flocks of sheep, straying wild and unguarded round about the fields and meadows; whereof they killed a good number, which they brought home with them in their ships. By which it appears, that the Norway Christians in the Western district were destroyed, and Christianity rooted out by the savage Heathens. The modern inhabitants of West Greenland, being, no doubt, the offspring of the before mentioned wild and barbarous Schrellings, have no certain account to give us of this matter; though they will tell you, that the old decayed dwelling places and villages, whose ruins are yet seen, were inhabited formerly by a nation quite different from theirs; and they also affirm, what the ancient histories tell us, that their ancestors made war with them, and destroyed them.[23]

---

[23] The Greenlanders relate a very ridiculous story, as well concerning the origin of our colonies (whom they call by the name of Kablunæt) as of their total overthrow, as follows: a Greenland woman, in her child-bearing, was once delivered of Kablunæt and dogs' whelps, of which the parents were highly ashamed, and for that reason withdrew from their neighbours and countrymen. This monstrous breed being grown up, became so troublesome to their father, that he was not able to endure them; wherefore he retired yet farther to some distant place. Meanwhile this inhuman race came to this horrible agreement amongst themselves, to devour their own father, whenever he should happen to come among them;

Now, as to the Eastern district, its present state is entirely unknown to us, as there is no approaching it with any shipping, upon account of the vast quantity of ice, driven from Spitzbergen and other Northern coasts upon this shore, which, adhering to the shore, barricades the land, and renders it wholly inaccessible. We may nevertheless gather from the above-mentioned expedition of the East Greenlanders against the Schrellingers, that after the destruction and total overthrow of the Western district and its colonies, the Eastern were yet standing and flourishing. But in what year this happened no notice is taken by the old historians. Nevertheless, from many tokens and remainders of probable evidence it may be inferred, that the old colony of the Eastern district is not yet quite extinct. To the confirmation of which, Thormoder, in his History of Greenland, alledges the following passage:—

Bishop Amand, of Shalholt in Iceland (who, anno 1522, had been consecrated, but, anno 1540, again resigned), once returning from Norway to Iceland, was by a storm driven Westward upon the coast of Greenland, which he coasted for some time Northwards, and made land towards the evening, finding themselves off Herjolsness. They came so near to the shore, that they could descry the inhabitants driving their flocks in the pasture grounds: but as the wind soon after proved fair they made all the sail they could, steering for Iceland, which they reached the day following, and entered the Bay of St. Patrick, which lies on the West coast of the island, in the morning early, when they were milking their cows.

Birn of Skarsaa (as we learn by the aforesaid Thormoder Torfager) gives the following relation:—

---

which a little after came to pass, when he visited them with a present of some part of a seal, which he had taken, according to custom. Kablunæt immediately went down to him, to whom the father delivered the piece of seal's flesh he had brought them. But he was no sooner got ashore, than the doggish race seized and devoured him, and then ate the seal's flesh given them. Whilst the Kablunæt dwelled there, one of the Innuits (or mankind), for so they call themselves, came rowing along the shore, and throwing his dart at some sea fowl, missed what he aimed at; which one of the Kablunæt, who stood upon a point of land running out into the sea, observing, mocked and ridiculed him, and, laying himself down upon the ground, told him that as he saw he was so dexterous in shooting, he would be the bird; he might throw the dart at him, and take care not to miss him: whereupon Innuit shot and killed him. This death caused continual strifes and wars between the Kablunæts and Innuits, which last at length became masters, and overthrew the former.

"In our time," says he, "one named John Greenlander, who for a considerable time had been employed in the service of the Hamburgh merchants, in a voyage from thence to Iceland, met with contrary winds and stormy weather, in which he narrowly escaped being cast away, and lost with ship and crew upon the dreadful rocks of Greenland, by getting in at last to a fine bay, which contained many islands, where he happily came to an anchor under a desert island; and it was not long before he spied several other islands not far off, that were inhabited; which, for fear of the inhabitants, he for a while did not dare to approach; till at last he took courage, and sending his boat on shore, went to the next house, which seemed but very small and mean. Here he found all the accoutrements necessary to fit out a fishing boat; he saw also a fishing booth, or small hut, made up of stones, to dry fish in, as is customary in Iceland. There lay a dead body of a man extended upon the ground with his face downwards; a cap sewed together on his head; the rest of his clothing was made partly of coarse cloth, and partly of seal skin; an old rusty knife was found at his side, which the captain took, in order to show it to his friends at his return home to Iceland, to serve for a token of what he had seen. It is farther said, that this commander was three times by stress of weather driven upon the coasts of Greenland, by which he obtained the surname of Greenlander."

This relation can be of no more than a hundred years standing, as Theodore Torlack affirms: because the above mentioned annals, in which we read it, were composed by Biorno of Skarsaa within these thirty years.

The same author furthermore informs us, that in Iceland there has often been found, scattered here and there on the sea shore, old broken pieces of deal boards, parts of the ribs of boats, which on the side were tacked together, and pasted with a sort of pitch or glue made of the blubber of seals. Now it is admitted, that this kind of glue is nowhere made use of but in Greenland; and a boat of this make was in the year 1625 found thrown up, upon a point of land near Reiche Strand, the structure of which was very artificial, joined together with wooden nails, not unlike that in which Asmund Kastenrazius, in the year 1189, in company with twelve men, crossed over from Greenland to Iceland; which boat was likewise tacked together with wooden nails, and the sinews of animals. The same historian,

in his book De Novitiis Groenlandorum Indiciis, tell us, that some years
ago, they found an oar upon the Eastern shore of Iceland, whereon these
words were carved in Runick characters: *Oft var ek dascedar ek dro dik*,
which signifies, "Often was I tired, when I carried thee." Besides this, I
find a relation in a German writer, whose name is Dithmarus Blefkenius,
concerning a certain monk, born in Greenland, who, as companion to the
bishop of the place, in the year 1546 made a voyage into Norway, where he
lived until the year 1564, and where, the author says, he got acquainted and
personally conversed with him. This monk told him many strange and
surprising things of a Dominican convent in Greenland, called St.
Thomas's Convent; to which his parents sent him in his youth to become a
monk of that order. But the truth of this relation is very much questioned,
being, together with several others of Blefkenius's relations, refuted and
gainsaid by Arngrim, in his Treatise, entitled Anatome Blefkeniana.
Blefkenius's relation is nevertheless confirmed by several other authors.
Erasmus Franciscus, in his book called East and West India State Garden,
in a place where he treats of Greenland tells us, that a captain of a Danish
ship, by name Jacob Hall, being ordered by the King his master to
undertake a voyage to Greenland, he touched first at Iceland, where he
from the King's lieutenant got intelligence of Greenland, which before was
unknown to him. And that he might the more fully be informed of
everything relating to this matter, a certain monk was sent for to instruct
him herein, who was said to be a native of Greenland; of whom the said
Jacob Hall, in his short description, gives the following account, according
to our above-mentioned author, Erasmus Fransciscus.

"There has formerly been a convent in Iceland, called Helgafield, or
Holy Mountain, in which, though it was decayed, lived a certain friar,
native of Greenland, with a broad and tawny face. This friar was sent for
by the King's lieutenant, in the presence of Jacob Hall, who wanted to be
informed of the state of Greenland. The friar accordingly told him, that
being very young, he was entered into this convent by his parents; and that
he afterwards was commanded by the same bishop, of whom he had
received the holy orders, to go along with him from thence to Norway,
where he submitted himself to the bishop of Drontheim, to whose authority

and jurisdiction all the priests of Iceland were subject; and being returned to his native home, he again retired and shut himself up in his former convent. This is said to have happened in the year 1546. He said moreover, that in the convent of St. Thomas, where he also had passed some time, there was a well of burning hot water, which, through pipes, was conveyed into all the rooms and cells of the convent to warm them."

But I think there is as much reason to question the authenticity of this relation as of the former, inasmuch as there is no such thing to be found in our Danish archives or annals. Notwithstanding which, what concerns St. Thomas's convent in particular is confessed, and confirmed by the old histories of Greenland. Nicolas Zenetur, a Venetian by birth, who served the King of Denmark in the quality of a sea captain, is said by chance to have been driven upon the coast of Greenland in the year 1380; and to have seen that same Dominican convent. His relation is alledged by Kircherus in the following words:—

"Here is also a Dominican convent to be seen, dedicated to St. Thomas, in whose neighbourhood there is a volcano of a mountain that vomits fire, and at the foot thereof a well of burning hot water. This hot water is not only conveyed by pipes into the convent, and through all the cells of the friars to keep them warm, as with us the rooms are heated by stoves of wood fire or other fuel; but here they also boil and bake their meat and bread with the same. This volcano, or fiery mountain, throws out such a quantity of pumice stone, that it hath furnished materials for the construction of the whole convent. There are also fine gardens, which reap great benefit from this hot water, adorned with all sorts of flowers, and full of fruit. And after the river has watered these gardens, it empties itself into the adjoining bay, which causes it never to freeze, and great numbers of fish and sea fowl flock thither, which yields plentiful provision for the nourishment of the inhabitants."

Of all the attested relations, that of Biorno of Skarsaa, concerning Bishop Amund of Skalholt, who was driven upon the coast of Greenland, deserves most to be credited; by which we learn, that the colony of the Eastern district flourished about one hundred and fifty years after the commerce and navigation ceased between Norway and Greenland; and, for aught we know, is not yet wholly destitute of its old Norwegian inhabitants. We have not been able to get any account of this matter from the modern Greenlanders, as they entertain no correspondence with those parts: either being hindered by the ice, which renders them altogether inaccessible; or else for fear the inhabitants of that country might kill and devour them; for they represent them as a cruel, barbarous, and inhuman nation, that destroy and eat all foreigners that fall into their hands. Yet notwithstanding this, if we may believe the relation of those adventurers, who have coasted a great part of the Eastern shore, there is no other sort of inhabitants found on this than on the Western side. But how it comes to pass, that the Eastern district, which was so well settled with Norway and Iceland colonies, that it contained twelve large parishes, and one hundred and ninety villages, besides one bishop's see and two convents, and flourished till the year 1540, at last has been destroyed and laid waste, is what I cannot conceive. The opinion of some, that the black plague, so called, which ravaged the Northern countries in the year 1348, also reached Greenland, and made its havock among its Eastern colonies, is without any ground or reason; because the commerce was carried into Greenland until the year 1406; and in 1540 that colony was still subsisting. If therefore this district be destitute or bereft of its old inhabitants, it is not unlikely they have undergone the same fatality as the Western ones, being destroyed by the barbarity of the savage Schrellingers.

A whole century passed from the cessation of all commerce and navigation between Norway and Greenland, till new adventurers began to apply themselves to the discovery of the Eastern district. The first of those who took this affair to heart was Erick Walkendorff, archbishop of Drontheim, who was resolved, at his own charge, to fit out ships for this purpose, but was stopped in this pious design by King Christian the Second, whose disgrace he had incurred. The next was King Frederick the First, whose mind, as it is reported, was bent upon the said expedition, but it was never put in execution. Christian the Third (as Lyscander relates) sent several ships with the same design, but without making any discovery. Frederick the Second succeeded his royal father, as well in the government as in his good design about Greenland; on which errand he sent Mogens

Heinson, a renowned seaman in those days. This adventurer, after he had gone through many difficulties and dangers of storms and ice, got sight of the land, but could not approach it; whereupon he returned home again, and pretended, that he might have got on shore, if his ship had not been stopped in the midst of its course, by some loadstone rocks hidden in the sea, so that he could not proceed though he had a very favourable and strong gale of wind, and no ice to hinder him: which frightened him and made him sail back again to Denmark. But the true loadstone rocks, in my opinion, was the terrible fright he was in of not getting safe through the dreadful ice mountains, which threatened him, or else the strong current, which always runs along the states promontory with such violence and rapidity, that it often stops a ship under full sail, so that the ship can make but little or no way at all against it. The cause by others assigned for this strange effect, the fish Remora, which the Northlanders call Kracken, is nothing but a fabulous story of the too credulous ancients, and labours under no less absurdities than the former opinion, that rocks of loadstone, laying on the bottom of the sea, can stay the course of a ship that sails on the surface of it.

In the same year that Mogens Heinson went upon the Greenland discovery, the English histories inform us, that Captain Martin Frobisher, an Englishman, was by the glorious Queen Elizabeth sent upon the same errand. This adventurer got sight of the land, but being partly hindered by the ice, which adhered to it, and partly by the shortness of the winter days (for it was late in the year), he could not approach it, and so returned to England again. Next year in the spring, he went upon the same expedition with three ships. After having gone through many great dangers of the ice and storms, he at length reached the shore, where he found a wild and savage nation; who, when they saw the English coming to them, being frightened, left their huts, and ran away to hide themselves. Some from the highest rocks threw themselves into the sea; whereupon the English entered their huts, where they met with nobody but an old woman, and a young one, who was pregnant, and those they carried away with them. It is also reported, that they here found some sand which contained particles of gold and silver, of which they filled three hundred tons, and brought it home with them to England. As to this gold and silver sand, I cannot help questioning whether they found any such on the Greenland shore, inasmuch as Sir Martin, in the same strain, relates wonderful things of the politeness and civility of a nation that dwelt in those parts; of which he says, they were governed by a prince, whom they called Kakiunge; and carried him in state on their shoulders, clothed in rich stuffs, and adorned with gold and precious stones, which does not at all agree with the meanness and coarseness of Greenland and its inhabitants; but rather seems to belong to the rich kingdoms of Peru and Mexico, where gold and silver abounds; and from whence he may have brought the above-mentioned gold and silver sand.

But I think it high time to leave such uncertain relations to their worth; and turn our thoughts towards the pious endeavours of our most gracious sovereigns the Kings of Denmark to discover and recover Greenland again. And we find, that after the expeditions of Frederick the Second, Christian the Fourth, his successor, with great cost, ordered four different expeditions for this discovery. The first was undertaken, under the command of Godske Lindenow, with three ships. And, as the history tells, Lindenow with his ship arrived upon the East coast of Greenland (which I hardly can believe), and found none but wild, uncivilised people there, like those Frobisher is said first to have met with. He stayed there three days, during which time the wild Greenlanders came to trade with him; changing all sorts of furs and skins with pieces of precious horns, against all kinds of small trifling iron ware, as knives, scissors, needles, common looking glasses, and other such trifles. When he set sail from thence, there were two Greenlanders remaining in the ship, whom he carried off, and brought them home along with him: these as they made all their endeavour to get away from him, and sometimes would have jumped into the sea, they were obliged to tie and secure them; which, when their countrymen observed, who flocked together upon the shore, they made a hideous outcry and howling, flung stones, and shot their arrows at the sailors, upon which they from the ship fired a gun, which frightened and dispersed them; and so the ship left them. The two other ships, that set sail in company and under the command of Lindenow, after they had doubled Cape Farewell, steered directly for the Strait of Davis; in which navigation they discovered many fine harbours and delightful green meadow lands, but all the inhabitants along the coast wild and savage as before. It is pretended also, that they in some places found stones, which contained some silver ore, which they took along with them; of which one hundred pounds yielded twenty-six ounces of silver. (Here again I cannot forbear questioning, whether this silver ore has been found on the Greenland shore, or rather over against it on the American coast.) These two ships also brought four savages home with them to Copenhagen.

The second expedition was made by order of the same King in the year 1606, with five ships under the conduct of the before-mentioned Admiral Lindenow; bringing along with them three of the savages (one of them dying in the voyage) which they had carried off the year before from Greenland. But this time he directed his course to the Westward of Cape Farewell, standing for the Straits of Davis; where he, coasting along, took the survey of several places, and then returned home again.

The third and last expedition of this glorious King was only of two ships, commanded by Captain Carsten Richards, a Holstenian by birth; he spied the land and its high and craggy rocks afar off, but could not come near it on account of the ice; and so, after he had lost his labour he returned home.

The fourth expedition of King Christian the Fourth, under the conduct of Captain Jens Munck, in the year 1616, was not made for the discovering of Greenland but to find out a passage between Greenland and America to China; the misfortunes of which expedition are related by the said commander.

There was, besides these four expeditions at the King's cost, a fifth undertaken, in the same King's reign, by a company settled in Copenhagen in the year 1636, of which company the president was the lord high chancellor, Christian Friis, as Lyscander informs us. Two ships fitted out by this company, directing their course to the Westward of Greenland, fell in with the Straits of Davis, where they traded for a while with the savages; but this was not the main concern of the commander, who was acquainted with a coast, whose sand had the colour and weight of gold, which he accordingly did not miss, and filled both their ships with the same. After their return to Copenhagen, the goldsmiths were ordered to make a trial, whether this sand would yield any gold or not; who, not being skillful enough to make such a trial, condemned it to be all thrown overboard, which was done by order of the high chancellor, president of the company. Some part of the said sand was yet kept out of curiosity, out of which an artificer, who afterwards came to Copenhagen, did extract a good deal of pure gold. The honest and well-meaning commander, who went upon this adventure, was turned out of favour, and died of grief soon after; whereby, not only the treasure they had brought home, but also the knowledge of the place where it was to be found, was entirely lost, as he kept this a secret to himself.

In the year 1654, during the reign of King Frederick the Third, a noble and wealthy adventurer, by name Henry Muller, fitted out a ship for Greenland, under the command of David de Nelles, who arrived safe in Greenland, and brought from thence three women, whose names were Kunelik, Kabelau, and Sigokou; who, according to the opinion of Bishop Torlais, who had perused the said captain's journal, were taken in the neighbourhood of Herjolsness, on the Eastern shore, as Thormoder Torfæus pretends; but which I cannot be made to believe. My opinion is, they were brought from the Western shore, near Baal's River, as some of the inhabitants, who are still living, had in fresh remembrance, telling me their names, as they are laid down in the fore-mentioned Journal.

The last adventurer, that was sent upon the discovery of Greenland, according to Torfæus in his History of Greenland, was Captain Otto Axelson, in the year 1670, in the reign of Christian V of glorious memory. But what success this adventurer met with he leaves us to guess. Nevertheless we find, in a manuscript description of Greenland, written by Arngrim Vidalin, Part iii, chap. 1, that his said majesty did invite, and with great privileges encourage Mr. George Tormúhlen, counsellor of commerce at Bergen, to fit out ships for the said discovery; whereupon the said counsellor not only got ready shipping well stored for such an expedition, but also got together a number of passengers, who resolved to go and settle in those parts, whom he provided with all things necessary for that purpose; both provision and ammunition, as well as houses made of timber, ready to be erected in that country. But this great design miscarried, the ship being taken by the French and brought into Dunkirk.

Thus, for a long while, it seemed, that all thought of Greenland was laid aside until the year 1721; when after many well-meant invitations, and projects proposed by me to the Greenland company at Bergen in Norway, approved and authorised by his late majesty Frederick IV of glorious memory, the company thereupon resolved not only to send ships, but also to settle a colony in Greenland in 64°; when I went over with my whole family and remained there fifteen years. During my stay I endeavoured to get all the intelligence that could be procured both by sea and land of the present state of the country, and did not lose my labour; for I found some places that formerly were inhabited by the old Norwegians, on the Western shore. Which expedition I have lately treated of in another treatise, and set out in all its circumstances, and with all the difficulties it has laboured under; wherefore I think it need not be here repeated.

But whereas my main drift and endeavour has been all along chiefly to discover the Eastern district of Greenland, which always was reckoned the best of our ancient colonies, accordingly I received from the above mentioned Greenland company at Bergen a letter, in the year 1723, in which I was told, that it was his majesty's pleasure, that the East district might likewise be visited and discovered. Which the better to effectuate, I took the resolution to make this voyage in person; and accordingly I coasted it Southwards, as far as to the States Promontory, looking out for the Strait of Frobisher, which would have been my shortest way, according to those charts, which lay the said strait down in this place; but such a strait I could not find. Now as it grew too late in the year for me to proceed farther, the month of September being nearly at an end, when the winter season begins in those parts, accompanied by dreadful storms, I was obliged to return.

In the year 1724 the directors of the said Bergen company, according to his majesty's good will and pleasure, fitted out a ship to attempt a landing on the Eastern shore, as had been formerly practised on that coast which lies opposite to Iceland. But the surprising quantity of ice, which

barricadoed the coast, made that enterprise prove abortive and quite
miscarry, as many others had done. As there was no appearance for ships
to approach this shore, the same king, in the year 1728, resolved, besides
other very considerable expenses, to have horses transported to this colony,
in hopes, that with their help they might travel by land to this Eastern
district: but nothing was more impossible than this, project, on account of
the impracticable, high, and craggy mountains perpetually covered with ice
and snow, which never thaws. Another new attempt by sea was by order of
the said king made in the year 1729, by Lieutenant Richard; who with his
ship passed the winter near the new Danish colony, in Greenland, and in
his voyage back to Denmark made all the endeavours he could to come at
the aforesaid shore, opposite to Iceland; but all to no purpose, being herein
disappointed, like the rest before him.

All these difficulties and continual disappointments have made most
people lose all hopes of succeeding in this attempt: nevertheless, I flatter
myself to have hit luckily on an expedient, which to me seems not
impracticable though hitherto not tried, or at least but lightly executed; *viz.*
to endeavour to coast the land from the States Promontory, or (as we call
it) Cape Prince Christian, Northwards. The information I have had of some
Greenlanders, who in their boats have coasted a great part of the East side,
confirms me in my opinion; for although an incredible quantity of driven
ice yearly comes from Spitzbergen or New Greenland along this coast, and
passes by the States Promontory, which hinders the approaching of ships as
far as the ice stretches, whereabout the best part of the Norwegian colonies
were settled; yet there have been found breaks and open sea near the shore,
through which boats and smaller vessels may pass; and according to the
relation of the Greenlanders, as well as agreeably to my own experience,
the current, that comes out of the bays and inlets, always running along the
shore South Westwards, hinders the ice from adhering to the land, and
keeps it at a distance from the shore; by which means the Greenlanders at
certain times, without any hindrance, have passed and repassed part of this
coast in their kone boats (so they call their large boats); though they have
not been so far as where the old Norway colonies had their settlement; of
which no doubt there are still some ruins to be seen on this Eastern shore.

Furthermore I have been credibly informed by Dutch seamen that frequent these seas, that several of their ships have at times found the East side of Greenland cleared of the ice as far as 62°; and they had tarried some time among the out rocks on that coast, where they carried on a profitable trade with the savages. And I myself, in my return from Greenland homewards in the year 1736, found it to be so when we passed the States Promontory and Cape Farewell, and stood in near the shore, where at that time there was no ice to be seen, which otherwise is very uncommon. But as this happens so seldom, it is very uncertain and unsafe for any ship to venture so far up under the Eastern shore. But, as I observed a little before, it is more safe and practicable to coast it from the Promontory along the shore in small vessels; especially if there be a lodge erected in the latitude of between 60° and 61°: and it would be still more convenient, if there could be a way and means found likewise to place a lodge on the Eastern shore in the same latitude. For according to the account the ancients have left us of Greenland, the distance of ground that lies uncultivated between the West and East side is but twelve Norway miles by water. See Ivarus Beri's relation; or, according to a later computation, it is a journey of six days in a boat. And as the ruins of old habitations, which I have discovered between 60° and 61°, are without doubt in the most Southerly part of the West side, it of necessity follows, that the distance cannot be very great from thence to the most Southern Parts of the Eastern side. Now, if it should be found practicable, at certain times, to pass along the shore with boats or small ships to the East side, to the latitude of 63° and 64°, little lodges might be settled here and there with colonies; by which means a constant correspondence might be kept, and mutual assistance given to one another, though larger ships could not yearly visit every one of them, but only touch at the most Southerly ones. I am also persuaded, that the thing is feasible, and if it should please God in his mercy to forward this affair, colonies might be established here, which, without great trouble, might be supplied yearly with all necessaries.

Chapter 3

TREATS OF THE NATURE OF THE SOIL, PLANTS, AND MINERALS OF GREENLAND

As to the nature of the soil, we are informed by ancient histories, that the Greenland colonies bred a number of cattle, which afforded them milk, butter, and cheese in such abundance, that a great quantity thereof was brought over to Norway, and for its prime and particular goodness was set apart for the King's kitchen, which was practised until the reign of Queen Margaret. We also read in these histories, that some parts of the country yielded the choicest wheat corn, and in the dales or valleys the oak trees

brought forth acorns of the bigness of an apple, very good to eat.[24] The
woods afforded plenty of game of rein deer, hares, &c. for the sport of
huntsmen. The rivers, bays, and the seas furnished an infinite number of
fishes, seals, morses, and whales; of which all the inhabitants make a
considerable trade and commerce. And though the country at present
cannot boast of the same plenty and richness, as it lies destitute of colonies,
cattle, and uncultivated; yet I do not doubt, but the old dwelling places,
formerly inhabited and manured by the ancient Norway colonies, might
recover their former fertility, if they were again peopled with men and
cattle; inasmuch as about those places there grows fine grass, especially
from 60° to 65°. In the great Bay, which in the sea charts goes under the
name of Baal's River, and at present is called the Bay of Good Hope (from
the Danish colony settled near the entrance of this inlet), there are on both
sides of the colony many good pieces of meadow ground, for the grazing
and pasturing numbers of cattle, besides plenty of provision, which the sea
as well as the land yields. Trees or woods of any consideration are rarely
met with; yet I have found in most of the bays underwoods and shrubs in
great quantity, especially of birch, elm, and willows, which afford
sufficient fuel for the use of the inhabitants, The largest wood I have seen
is in the latitude of 60° and 61°, where I found birch trees two or three
fathom high, somewhat thicker than a man's leg or arm: small juniper trees
grow also here in abundance, the berries of which are of the bigness of
grey peas. The herb called quaun, which is our angelica, is very obvious
and common, as well as wild rosemary, which has the taste and smell of
turpentine; of which, by distillation, is extracted a fine oil and spirit, of
great use in medicine. That precious herb, scurvy grass, the most excellent
remedy for the cure of the distemper which gives its name, grows
everywhere on the sea side, and has not so bitter a taste as that of softer
climates; I have seen wonderful effects of its cure. The country also
produces a grass with yellow flowers, whose root smells in the spring like

---

[24] A Greenlander, who came from the most Southern part of the country near the States
Promontory, told my son, when he saw some lemons in his room, that he had seen fruits
much like those growing upon trees in his country, though they were four times less; which
I take to have been some of those acorns, which I above took notice of, treating of the
nature of the soil.

roses: the inhabitants feed thereupon, and find benefit by it. In the bays and inlets you have wild thyme at the side of the mountains, which after sunset yields a fragrant smell. Here also you meet with the herb tormentil, or setfoil, and a great many other herbs, plants, and vegetables, which I cannot call to mind, and whose names indeed are altogether unknown to me. Their most common berries are those called blew-berries, tittle-berries, and bramble-berries. Multe-berries, which are common in Norway, do not arrive here to any perfection, on account of the thick fogs that hang upon the islands, when these plants bud. This country affords the most pleasant prospect about the latitude of 60° to 64°, and seems fit to be manured for the produce of all sorts of grain; and there are to this day marks of acres and arable land to be observed. I myself once made a trial of sowing barley in the bay joining to our new colony, which sprung up so fast, that it stood in its full ears towards the latter end of July; but did not come to ripeness, on account of the night frost which nipped it and hindered its growth. But as this grain was brought over from Bergen in Norway, no doubt it wanted a longer summer and more heat to ripen. But I am of opinion, that corn which grows in the more Northern parts of Norway would thrive better in Greenland, inasmuch as those climates agree better together. Turnips and cole are very good here, and of a sweet taste, especially the turnips, which are pretty large.

I must observe to you, that all that has been said of the fruitfulness of the Greenland soil is to be understood of the latitude of 60° to 65°, and differs according to the different degrees of latitude. For in the most Northern parts you find neither herbs nor plants; so that the inhabitants cannot gather grass enough to put in their shoes to keep their feet warm, but are obliged to buy it from the Southern parts.

Of Greenland metals or minerals I have little or nothing to say. It is true, that about two Norway miles to the South of the colony of Good Hope, on a promontory, there are here and there green spots to be seen, like verdigris, which shows there must be some copper ore. And a certain Greenlander once brought me some pieces not unlike lead ore. There is likewise a sort of calamine, which has the colour of yellow brass. In my expedition upon discoveries, I found, on a little island where we touched, some yellow sand, mixed with sinople red, or vermillion strokes, of which I sent a quantity over to the directors of the Greenland company at Bergen, to make a trial of it; upon which they wrote me an answer, that I should endeavour to get as much as I could of the same sand; but to theirs as well as my own disappointment, I never was able to find the said island again, where I had got this sand, as it was but a very small and insignificant one, situate among a great many others; and the mark I had taken care to put up was by the wind blown down. Nevertheless there has been enough of the same stuff found up and down in the country, which, when it is burnt, changes its former colour for a reddish hue, which it likewise does if you keep it awhile shut up close.

Whether or not this be the same sort of sand as that of which Sir Martin Frobisher is said to have brought some hundred tons to England, and was pretended to contain a great deal of gold; and again (as we have above taken notice of) of which some of the Danish Greenland Company's ships returned freighted to Copenhagen in the year 1636, is a question which I have no mind to decide. However, thus much I can say, that by the small experience I have acquired in the art of chemistry, I have tried both by extraction and precipitation if it would yield anything, but always lost my labour. After all I declare, I never could find any other sort of sand that contained either gold or silver. But as for rock crystal, both red and white, you find it here: the red contains some particular solis, which can only be produced by the spagyric art.

Stone flax, or what they call asbestos, is so common here, that you may see whole mountains of it: it has the appearance of a common stone, but can be split or cloven like a piece of wood. It contains long filaments, which, when beaten and separated from the dross, you may twist and spin into a thread. As long as it has its oily moisture it will burn without being consumed to ashes.

Round about our colony of Good Hope there is a sort of coarse bastard marble of different colours, blue, green, red, and some quite white, and again some white with black spots, which the natives form into all sorts of vessels and utensils, as lamps, pots to boil in, and even crucibles to melt metals in, this marble standing proof against the fire.[25] Of this marble there was brought a quantity over to Drontheim in Norway, which they made use of in the adorning of the cathedral of that city, as we have it from Peter Claudius Undalin[26].

Amongst the produce of the sea, besides different shells, muscles, and periwinkles, there are also coral trees, of which I have seen one of a fine form and size.

---

[25] The lamps and pots, which the Southern Greenlanders make of this marble, are sold at a very high price; so that the natives of the Northern parts, where such marble is not to be had, buy them at the rate of eight or ten rein deer skins a large pot, and a lamp at two or three skins.

[26] According to what the natives tell, there is in the Southern parts a hot well, of a mineral quality; which, if you wash therein, cures the itch; they wash their skins in it, and it takes away all dirt and foulness, and makes them look like new.

Chapter 4

OF THE NATURE OF THE CLIMATE, AND THE TEMPERAMENT OF THE AIR

The natives of Greenland have no reason to complain of rains and stormy weather, which seldom trouble them; especially in the Bay of Disco, in the 68th degree of Latitude, where they commonly have clear and settled weather during the whole summer season: but again, when foul and stormy weather falls in, it rages with an incredible fierceness and violence, chiefly when the wind comes about Southerly, or South West; and the storm is laid and succeeded by fair weather as soon as the wind shifts about to the West and North.

The country would be exceeding pleasant and healthful in summer time, if it was not for the heavy fogs that annoy it, especially near the sea coast; for it is as warm here as anywhere, when the air is serene and clear, which happens when the wind blows Easterly; and sometimes it is so hot, that the sea water, which after the ebbing of the sea has remained in the hollow places of the rocks, has often, before night, by the heat of the sun, been found coagulated into a fine white salt. I can remember, that once, for three months together, we had as fair settled weather and warm sunshine days as one could wish, without any rain.

The length of the summer is from the latter end of May to the midst of September; all the remaining part of the year is winter, which is tolerable in the latitude of 64°, but to the Northward, in 68° and above, the cold is so excessive, that even the most spirituous liquors, as French brandy, will freeze near the fire side. At the end of August the sea is all covered with ice, which does not thaw before April or May, and sometimes not till the latter end of June.

It is remarkable, that on the Western coasts of different countries, lying in one and the same latitude, it is much colder than on the Eastern, as some parts of Greenland and Norway. And though Greenland is much colder than Norway, yet the snow never lies so high, especially in the bays and inlets, where it is seldom above half a yard higher than the ground; whereas the inland parts and the mountains are perpetually covered with ice and snow, which never melts; and not a spot of the ground is bare, but near the shore and in the bays; where in the summer you are delighted with a charming verdure, caused by the heat of the sun, reverberated from side to side, and concentred in these lower parts of the valleys, surrounded by high rocks and mountains, for many hours together without intermission; but as soon as the sun is set, the air is changed at once, and the cold ice mountains make you soon feel the nearness of their neighbourhood, and oblige you to put on your furs. Besides the frightful ice that covers the whole face of the land, the sea is almost choaked with it, some flat and large fields of ice, or bay ice, as they call it, and some huge and prodigious mountains, of an astonishing bigness, lying as deep under water as they soar high in the air. These are pieces of the ice mountains of the land, which lie near the sea, and bursting, tumble down into the sea, and are carried off. They represent to the beholders, afar off, many odd and strange figures; some of churches, castles with spires and turrets others you would take to be ships under sail; and many have been deluded by them, thinking they were real ships, and going to board them. Nor does their figure and shape alone surprise, but also their diversity of colours pleases the sight; for some are like white crystal, others blue as sapphires; and others again green as emeralds. One would attribute the cause of these colours to metals or minerals of the places where this ice was formed; or of waters of which it was coagulated: but experience teaches me, that the blue ice is the concretion of fresh water, which at first is white, and at length hardens and turns blue; but the greenish colour comes from salt water. It is observed, that if you put the blue ice near the fire and let it melt, and afterwards remove it to a colder place, to freeze again, it does not recover its former blue, but becomes white. From whence I infer, that the volatile sulphur, which the ice had attracted from the air, by its resolution into water, exhales and vanishes.

Though the summer season is very hot in Greenland, it seldom causes any thunder and lightning; the reason of which I take to be the coolness of the night, which allays the heat of the day, and causes the sulphureous exhalations to fall again with the heavy dew to the ground.

As for the ordinary meteors, commonly seen in other countries, they are visible in Greenland; as the rainbow, flying or shooting stars, and the like. But what is more peculiar to the climate, is the Northern Light, or Aurora Borealis, which in the spring of the year, about the new moon, darts streams of light all over the sky, as quick as lightning, especially if it be a clear night, with such a brightness, that you may read by it as by daylight.

At the summer solstice there is no night, and you have the pleasure to see the sun turn round about the horizon all the twenty-four hours; and in the depth of winter they have but little comfort in that planet, and the nights are proportionally long; yet it never is so dark, but you can see to travel up and down the country, though sometimes it be neither moonshine nor starlight: but the snow and ice, with which both land and sea is covered, enlightens the air; or the reason may be fetched from the nearness of the horizon to the equator.

The temperament of the air is not unhealthful; for, if you except the scurvy and distempers of the breast, they know nothing here of the many other diseases with which other countries are plagued; and these pectoral infirmities are not so much the effects of the excessive cold, as of that nasty foggish weather which this country is very much subject to; which I impute to the vast quantity of ice that covers the land and drives in the sea. From the beginning of April to the end of July is the foggish season, and from that time the fog daily decreases. But as in the summer time they are troubled with the fog, so in the winter season they are likewise plagued with the vapour called frost smoke, which, when the cold is excessive, rises out of the sea as the smoke out of a chimney, and is as thick as the thickest mist, especially in the bays, where there is any opening in the ice. It is very remarkable, that this frost, damp, or smoke, if you come near it, will singe the very skin of your face and hands; but when you are in it, you find no such piercing or stinging sharpness, but warm and soft; only it leaves a white frost upon your hair and clothes.

I must not forget here to mention the wonderful harmony and correspondence which is observed in Greenland between fountains and the main sea, *viz.* that at spring tides, in new and full moon, when the strongest ebbing is at sea, the hidden fountains or springs of fresh water break out on shore, and discover themselves, often in places where you never would expect to meet with any such; especially in winter, when the ground is covered with ice and snow; yet at other times there are no water springs in those places. The cause of this wonderful harmony I leave to the learned inquiry of natural philosophers; how springs and fountains follow the motion of the main sea, as the sea does that of the moon. Yet this I must observe to you, that some great men have been greatly mistaken, in that they have taken for granted and asserted, that in Norway and Greenland the tide was hardly remarkable. (See Mr. Wollf's Reasonable Thoughts on the Effects of Nature, p. 541.) Whereas nowhere greater tide is observed; the sea, at new and full moon, especially in the spring and fall, rises and falls about three fathoms.

Chapter 5

OF THE LAND ANIMALS, AND LAND FOWLS OR BIRDS OF GREENLAND; AND HOW THEY HUNT AND KILL THEM

There are no venomous serpents or insects, no ravenous wild beasts to be seen in Greenland, if you except the bear, which some will have to be an amphibious animal, as he lives chiefly upon the ice in the most Northern parts, and feeds upon seals and fish. He very seldom appears near the colony, in which I had taken up my quarters. He is of a very large size, and of a hideous and frightful aspect, with white long hairs: he is greedy of human blood.[27] The natives tell us moreover of another kind of ravenous beasts, which they call Amarok, which eagerly pursue other beasts, as well as men; yet none of them could say, they ever had seen them, but only had it from others by hearsay; and whereas none of our own people, who have travelled up and down the country, ever met with any such beast, therefore I take it to be a mere fable.

Rein deer are in some places in so great numbers that you will see whole herds of them;[28] and when they go and feed in herds they are dangerous to come at. The natives spend the whole summer season in hunting of rein deer, going up to the innermost parts of the bays, and carrying, for the most part, their wives and children along with them, where they remain till the harvest season comes on. In the mean while they with so much eagerness hunt, pursue, and destroy these poor deer, that they have no place of safety, but what the Greenlanders know; and where they are in any number, there they chase them by clap-hunting, setting upon

---

[27] In the 76th degree of latitude the number of bears is so great, that they in droves surround the natives' habitations, who then, with their dogs, fall upon them, and with their spears and lances kill them. In winter, instead of dens or caves under the earth, as in Norway and other places, here the bears make theirs under the snow; which, according to the information the natives have given me, are made with pillars, like stately buildings.

[28] The farther you go Northwards, the seldomer you meet with rein deer, except in the 3d or 4th degree to the North of Disco, where they are in great numbers; perhaps by reason either of its joining to America, or else because the deer pass over to the islands upon the ice, in quest of food, which the main land, covered with ice and snow, does not afford them. The natives, instead of reason, give us a very childish tale for the vast number of rein deer being found upon Disco Island, as follows:—

A mighty Greenlander (one Torngarsuk, as they call him, who is father to an ugly frightful woman, who resides in the lowermost region of the Earth, and has command over all the animals of the sea, as we shall see hereafter) did with his Kajar, tow this island to the place where it now lies, from the South where it was before. Now, as the face of this island resembles very much the Southern coasts, and the root angelica is likewise found upon it, which grows nowhere else in the neighbouring parts, this confirms them in their credulity. And furthermore, they assure you, that a hole is seen to this day in the island, through which the towing-rope had been fastened by Torngarsuk.

them on all sides, and surrounding them with all their women and children, to force them into defiles and narrow passages, where the men armed lay in wait for them and kill them: and when they have not people enough to surround them, then they put up white poles (to make up the number that is wanted) with pieces of turf to head them, which frightens the deer, and hinders it from escaping.

There are also vast numbers of hares, which are white summer and winter, very fat and of a good taste. There are foxes of different colours, white, grey, and blueish; they are of a lesser size than those of Denmark and Norway, and not so hairy, but more like martens. The natives commonly catch them alive in traps, built of stones like little huts. The other four-footed animals, which ancient historians tell us are found in Greenland, are sables, martens, wolves, losses, ermins, and several others; I have met with none of them on the Western side.—See Arngrim Jonas's History of Greenland; as also Ivarus Beni's Relation, mentioned by Undalinus.

Tame or domestic animals there are none, but dogs in great numbers, and of a large size, with white hairs, or white and black, and standing ears. They are in their kind as timorous and stupid as their masters, for they never bay or bark, but howl only. In the Northern parts they use them instead of horses, to drag their sledges, tying four or six, and sometimes eight or ten to a sledge, laden with five or six of the largest seals, with the master sitting up himself, who drives as fast with them as we can do with good horses, for they often make fifteen German miles with them in a winter day, upon the ice: and though the poor dogs are of so great service to them, yet they do not use them well, for they are left to provide for and subsist themselves as wild beasts, feeding upon muscles thrown up on the sea side, or upon berries in the summer season; and when there has been a great capture of seals they give them their blood boiled and their entrails.

As for land fowls or birds, Greenland knows of none but rypper, which is a sort of large partridges, white in winter, and grey in summer time, and these they have in great numbers. Ravens seem to be domestic birds with them, for they are always seen about their huts, hovering about the carcases of seals, that lie upon the ground. There are likewise very large eagles, their wings spread out being a fathom wide, but they are seldom seen in the Northern parts of the country. You find here falcons or hawks, some grey, some of a whitish plumage, and some speckled; as also great speckled owls. There are different sorts of little sparrows, snow birds, and ice birds, and a little bird not unlike a linnet, which has a very melodious tune.

Amongst the insects of Greenland, the midge or gnats are the most troublesome, whose sting leaves a swelling and burning pain behind it; and this trouble they are most exposed to in the hot season, against which there is no shelter to be found. There are also spiders, flies, humble bees, and wasps. They know nothing of any venomous animals, as serpents and the like; nor have they any snakes, toads, frogs, beetles, ants, or bees; neither are they plagued with rats, mice, or any such vermin.

Chapter 6

OF THE GREENLAND SEA ANIMALS, AND SEA FOWLS AND FISHES

The Greenland Sea abounds in different sorts of animals, fowls, and fishes, of which the whale bears the sway, and is of divers kinds, shapes, and sizes. Some are called the finned whales, from the fins they have upon their back near the tail; but these are not much valued, yielding but little fat or blubber, and that of the meaner sort; they consist of nothing but lean flesh, sinews, and bones. They are of a long, round, and slender shape, very dangerous to meddle with, for they rage and lay about them most furiously with their tail, so that nobody cares to come at them, or catch them. The Greenlanders make much of them, on account of their flesh, which, with them, passes for dainty cheer. The other sort of whales are reckoned the best for their fat, and fins or whalebones. These differ from the first sort, in that they have no fin on the back towards the tail, but two lesser ones near the eyes, and are covered with a thick black skin, marbled with white strokes. With these side fins they swim with an incredible swiftness. The tail is commonly three or four fathoms broad. The head makes up one-third of the whole fish. The jaws are covered, both above and beneath, with a kind of short hair. At the bottom of the jaws are placed the so called barders, or whalebones, which serve him instead of teeth, of which he has none. They are of different colours, some brown, some black, and others yellow with white streaks. Within the mouth, the barders or whalebones are covered with hair like horse-hair, chiefly those that inclose the tongue. Some of them are bent like a scymitar, or sabre. The smallest are ranged the foremost in the mouth, and the hindermost near the throat; the broadest and largest are in the middle, some of them two fathoms long, by which we may judge of the vast bigness of this animal. On each side there are commonly two hundred and fifty, in all five hundred pieces. They are set in a broad row, as in a sheaf, one close to the other, bent like a crescent or half-moon, broadest at the root, which is of a tough and grisly matter, of a whitish colour, fastened to the upper part of the jaws near the throat, and they grow smaller towards the end, which is pointed; they are also covered with hair, that they may not hurt the tongue. The undermost jaw is commonly white, to which the tongue is fastened, inclosed in the barders, or long whale bones; it is very large, sometimes about eighteen feet, and sometimes more, of a white colour, with black spots, of a soft, fat, and spungy matter. The whale has a bunch on the top of his head, in which are two spouts or pipes, parallel one to the other, and somewhat bent, like the holes upon a fiddle. Through these he receives the air, and spouts out the water, which he takes in at his mouth, and is forced upwards through these holes in very large quantities, and with such violence and noise, that it is heard at a great distance, by which, in hazy weather, he is known to be near, especially when he finds himself wounded, for then he rages most furiously, and the noise of his spouting is so loud, that some have resembled it to the roaring of the sea in a storm, or the firing of great guns, His eyes are placed between the bunch and the side fins; they are not larger than those of an ox, and are armed with eyebrows.

The penis of a whale is a strong sinew, seven or eight, and sometimes fourteen feet long, in proportion to his bulk: it is covered with a sheath, in which it lies hidden, so that you see but little of it: the nature of the female is like that of the four-footed animals: she has two breasts with teats like a cow; some white, others stained with black or blue spots. In their spawning time their breasts are larger than usual; and when they couple together, they reach their head above water, to fetch breath, and to cool the heat contracted by that action. It is said, that they never bring forth more than two young ones at a spawning, which they suck with their teats. The spawn of the whale, while it is fresh, is clammy and gluish, so that it may be drawn out in threads like wax or pitch; it has no relation to that which we call spermaceti, for it is soon corrupted and by no art can be preserved.

These sea animals, or rather monsters, are of different sizes and bulks; some yield one hundred, and some two or three hundred tons of fat or blubber. The fat lies between the skin and the flesh, six or eight inches thick, especially upon the back and under the belly. The thickest and strongest sinews are in the tail, which serves him for a rudder, as his fins do for oars, wherewith he swims with an astonishing swiftness, proportioned to his bulk, leaving a track in the sea, like a great ship; and this is called his wake, by which he is often followed.

These sea monsters are as shy and timorous as they are huge and bulky, for as soon as they hear a boat rowing, and perceive any body's approach, they immediately shoot under water and plunge into the deep; but when they find themselves in danger, then they shew their great and surprising strength; for then they break to pieces whatever comes in their way, and if they should hit a boat, they would beat it in a thousand pieces. According to the relation of the whale-catchers, the whale, being struck, will run away with the line some hundreds of fathoms long, faster than a ship under full sail. Now one would think, that such a vast body should need many smaller fishes and sea animals to feed upon; but on the contrary, his food is nothing but a sort of blubber, called *pulmo marinus*, or whale food, which is of a dark brown colour, with two brims or flaps, with which it moves in the water, with such slowness that one may easily lay hold of it, and get it out of the water. It is like a jelly, soft and slippery, so that if you crush it between your fingers you find it fat and greasy like train oil. The Greenland seas abound in it, which allures and draws this kind of whales thither in search of it; for as their swallow or throat is very narrow (being but four inches in diameter), and the smaller whalebones reaching down his throat, they cannot swallow any hard or large piece of other food, having no teeth to chew it with, so that this sort of nourishment suits them best, their mouth being large and wide to receive a great quantity, by opening it and shutting it again, that nature has provided them with the barders or whalebones, which by their closeness only give passage to the water, like a sieve, keeping back the aliment. Here we ought to praise the wise and kind providence of an Almighty Creator, who has made such mean things suffice for the maintenance of so vast an animal.

Next to this there is another sort of whales, called the North Capers, from the place of their abode, which is about the North cape of Norway, though they also frequent the coasts of Iceland, Greenland, and sundry other seas, going in search of their prey, which is herring and other small fish, that resort in abundance to those coasts. It has been observed, that some of these North Cape whales have had more than a ton of herrings in their belly. This kind of whales has this common with the former called fin-whale, in that it is very swift and quick in its motion, and keeps off from the shore in the main sea, as fearing to become a prey to its enemies, if it should venture too near the shore. His fat is tougher and harder than that of the great bay whale; neither are his barders or bones so long and valuable, for which reason he is neglected.

The fourth sort is the sword-fish, so called from a long and broad bone, which grows out of the end of his snout on both sides, indented like a saw. He has got two fins upon his back, and four under the belly, on each side two: those on the back are the largest; those under the belly are placed just under the first of the back: his tail broad and flat underneath, and above pointed, but not split or cloved. From the hindermost fin of the back he grows smaller: his nostrils are of an oblong shape: the eyes are placed on the top of his head, just above his mouth. There are different sizes of sword-fish, some of twenty feet, some more, some less. This is the greatest enemy the true whale has to deal with, who gives him fierce battles; and, having vanquished and killed him, he contents himself with eating the tongue of the whale, leaving the rest of the huge carcase for the prey and spoils of the morses and sea birds.

The cachelot or pot-fish is a fifth species of whales, whose shape is somewhat different from that of other whales, in that the upper part of his head or skull is much bigger and stronger built; his spouts or pipes are placed on the forehead, whereas other whales have them on the hinder part of the head: his under jaw is armed with a row of teeth which are but short: his tongue is thin and pointed, and of a yellowish colour: he has but one eye on the side of the head, which makes him of easy access to the Greenlanders, who attack him on his blind side. Of his skull that wrongly so called spermaceti is prepared, one yielding twenty to twenty-four tuns thereof. The rest of the body and the tail are like unto those of other whales. He is of a brownish colour on the back, and white under the belly: he is of different sizes, from fifty to seventy feet long.

Then comes the white fish, whose shape is not unlike that of the great bay whale, having no fins upon the back, but underneath two large ones; the tail like a whale; his spouts, through which he breathes and throws out the water, are the same; he has likewise a bunch on the head: his colour is of a fading yellow; he is commonly from twelve to sixteen feet in length, and is exceeding fat. The train of his blubber is as clear as the clearest oil: his flesh as well as the fat has no bad taste, and when it is marinated with vinegar and salt, it is as well tasted as any pork whatsoever. The fins also and the tail, pickled or sauced, are good eating. This fish is so far from being shy, that whole droves are seen about the ships at sea: the Greenlanders catch numbers of them, of which they make grand cheer.

There is yet another smaller sort of whales, called but-heads, from the form of its head, which at the snout is flat, like a but's end: he has a fin upon his back towards the tail, and two side fins: his tail is like to that of a whale. In the hinder part of the head he has a pipe to fetch air, and spout the water through, which he does not spout out with that force the whale does: his size is from fourteen to twenty feet: he follows ships under sail with a fair wind, and seems to run for a wager with them; whereas, on the contrary, other whales avoid and fly from them. Their jumping, as well as that of fishes and sea animals, forebodes boisterous and stormy weather.

Among the different kinds of whales some reckon the unicorn, as they commonly call him, from a long small horn that grows out of his snout; but his right name is nar-whale. It is a pretty large fish, eighteen or twenty feet long, and yields good fat: his skin is black and smooth without hair; he has one fin on each side, at the beginning of his belly: his head is pointed, and out of his snout on the left side proceeds the horn, which is round, turned, with a sharp taper point; the greatest length of it is fourteen or fifteen feet, and thick as your arm. The root of it goes very deep into the head, to strengthen it for supporting so heavy a burthen. The horn is of a fine, white, and compact matter, wherefore it weighs much: the third part of it, beginning from the root, is commonly hollow; and there are some very solid at the root, and above it grows more and more hollow. On the right side of the head there lies another shorter horn hidden, which does not grow out of the skin, and it cannot be conceived for what end the All-wise Creator has ordained it: he has, like other whales, two pipes or spouts which terminate in one, through which he breathes and fetches air, when he comes up out of the sea with his head. Here I must observe to you, that when the whale comes up to fetch air, it is not water he throws out at the spouts, as the common notion runs; but his breath, which resembles water forced out of a great spout. As for the rest of the unicorn or nar-whale's body, it is perfectly of the same shape as that of other whales.

Concerning this animal's horn, which has given occasion to so many disputes, whether it be a horn properly so called, or a tooth, my reader must allow me a little digression, to make these gentlemen disputants aware of their mistake, who pretend it to be a tooth and not a horn, being placed on one side of the snout, and not on the top of the forehead, where other animals wear their horns. (See Wormius's Museum, l. iii. ch. 14.) But it appears clearly to all beholders, that it neither has the shape of a tooth, such as other sea animals are endowed with, nor has its root in the jaws, the ordinary place of teeth, but grows out of the snout. And besides, the absurdity is much greater to hold and maintain, that animals wear teeth on the snout or head, like horns: or dare anybody deny, that the whale's spouts are his nostrils, through which he fetches breath, because they are on the top of his head; or question, that the clap-mysses' (a large kind of seal) eyes are such, because they are placed in the hindermost part of the head? Ought we not rather to think, that an All-wise Creator has placed this horn horizontally, to the end that it may not be of any hinderance to the course and swimming of this animal in the water, which would happen if it rose vertically? Furthermore, this horn serves many other ends, as to stir up his food from the bottom of the sea, as he is said to feed upon small sea-weeds, and likewise therewith to bore holes in the ice, in order to fetch fresh air. The inference these gentlemen are pleased to draw from the generality of fishes and sea animals having no such paws or claws as land animals have, is as lame, and of as little force. And it is much less absurd to hold, that sea animals have something common with those of the land, as it is confessed, that many of them have a great resemblance together in figure and shape, viz. sea-calves, sea-dogs, sea-wolves, and sea-horses, together with mermen and mermaids, as it is pretended. Who is ignorant of the winged or flying fishes; and of others with long nebs or bills like birds; also of birds with four feet like beasts, and why then may there not be sea-unicorns as well as land unicorns; if any such there be in *rerum natura*? for it is a difficult matter to determine what kind of animal the Scripture understands, when it speaks of the unicorn, as in Psalm xxix. ver. 6, and in other places; whether it be such a one as Plinius and other writers describe, giving him the body of a horse, with a stag's head, and a horn on his snout; or whether it ought not with better reason be applied to a certain animal in Africa, called rhinoceros, whose snout is horned in that fashion. If one had patience to consider the vast disagreement that reigns between these writers, one would conclude that this animal is peculiar to the climate where the fabulous bird phœnix builds its nest; that is to say in Utopia, or nowhere. For some describe this animal as an amphibious one, that lives by turns upon land and in the water; some will have him to be in the likeness of an ore white spotted, with horse feet; others make a three years' colt of him, with a stag's head, and a horn in the front one ell long; and others again tell you it is like a morse or sea-horse, with divided or cloven feet, and a horn in the front. There are authors, who attribute to him a horn ten feet long, others six, and others again but the length of three inches. (See Pliny, Munsterus, Marc. Paulus, Philostratus, Heliodorus, and several others, whose relations are of the same authority with mine, as that of the Greenlanders, concerning a fierce, ravenous wild beast, which they call Amavok; which all pretend to know, but no person ever yet was found, that could say he had seen it.)

Nises or porpoises, otherwise sea hogs, are also placed in the class of whales, though of a much smaller size, and are met with in all seas. His head resembles that of a butts-head-whale: his mouth is armed with sharp teeth: he has spouts or pipes like a whale. He has a fin upon the middle of his back, which towards the tail is bended like a half-moon. Under the belly there are two side fins, overgrown with flesh and covered with a black skin. His tail is broad like that of a whale. He has small round eyes; his skin is of a shining black, and the belly white. His length is five to eight feet, at most. His fat makes fine oil, and the flesh is by the Greenlander reckoned a great dainty.

Of Other Sea Animals

The sea horse or morse has the shape of a seal, though much larger and stronger. He has five claws on each of his feet, as the seal: his head rounder and larger. His skin is an inch thick, especially about the neck, very rough, rugged and wrinkled, covered with a short, brown, and sometimes reddish, or mouse-coloured hair. Out of his upper jaw there grow two large teeth or tusks, bended downwards over the under jaw, of the length of half a yard, and sometimes of a whole yard and more. These tusks are esteemed as much as elephants' teeth; they are compact and solid, but hollow towards the root. His mouth is not unlike that of a bull, covered above and beneath with strong bristles as big as a straw: his nostrils are placed above his mouth, as those of the seal: his eyes are fiery red, which he can turn on all sides, not being able to turn his head, by reason of the shortness and thickness of his neck. The tail resembles a seal's tail, being thick and short: his fat is like hog's lard. He lies commonly upon the ice shoals, and can live a good while on shore, till hunger drives him back into the seas; his nourishment being both herbs and fishes: he snores very loud, when he sleeps; and when he is provoked to anger, he roars like a mad bull. It is a very bold and fierce creature, and they assist each other, when attacked, to the last. He is continually at war with the white bear, to whom he often proves too hard with his mighty tusks, and often kills him, or at least does not give over till they both expire.

The seals are of different sorts and sizes, though in their shape they all agree, excepting the clap-myss, so called from a sort of a cap he has on his head, with which he covers it when he fears a stroke. The paws of a seal have five claws, joined together with a thick skin, like that of a goose or a water fowl: his head resembles a dog's with cropped ears, from whence he has got the name of sea dog: his snout is bearded like that of a cat: his eyes are large and clear with hair about them: the skin is covered with a short hair of divers colours, and spotted; some white and black, others yellowish, others again reddish, and some of a mouse colour: his teeth are very sharp and pointed. Although he seems lamish behind, yet he makes nothing of getting up upon the ice hills, where he loves to sleep and to bask himself in the sun. The largest seals are from five to eight feet in length; their fat yields better train-oil than that of any other fish. This is the most common of all the sea animals in Greenland; and contributes the most to the subsisting and maintaining of the inhabitants, who feed upon the flesh of it, and clothe themselves with the skin, which likewise serves them for the covering of their boats and tents: the fat is their fuel, which they burn in their lamps, and also boil their victuals with. As for other sea monsters and wonderful animals, we find in Tormoder's History of Greenland, mention made of three sorts of monsters, where he quotes a book, called "Speculum Regale Iclandicum;" or, the Royal Island Looking-Glass, from whence he borrows what he relates.[29] But none of them have been seen by us, or any

---

[29] The above-mentioned author calls the first of these monsters Havestramb, or Mer-man, and describes it to have the likeness of a man, as to the head, face, nose, and mouth; save that its head was oblong and pointed like a sugar-loaf; it has broad shoulders, and two arms without

of our time, that ever I could hear, save that most dreadful monster, that
showed itself upon the surface of the water in the year 1734, off our new

hands; the body downwards is slanting and thin; the rest below the middle, being hid in the water, could not be observed. The second monster he calls Margya, or Mer-woman, or Mermaid, had from the middle upwards the shape and countenance of a woman; a terrible broad face, a pointed forehead, wrinkled cheeks, a wide mouth, large eyes, black untrimmed hair, and two great breasts, which showed her sex; she has two long arms, with hands and fingers joined together with a skin, like the feet of a goose; below the middle she is like a fish, with a tail and fins. The fishermen pretend, that when these sea monsters appear, it forebodes stormy weather. The third monster, named Hafgufa, is so terrible and frightful, that the author does not well know how to describe it; and no wonder, because he never had any true relation of it: its shape, length, and bulk, seems to exceed all size and measure. They that pretend to have seen it, say, it appeared to them more like a land than a fish, or sea animal. And as there never has been seen above two of them in the wide open sea, they conclude, that there can be no breed of them; for if they should breed and multiply, all the rest of fishes must be destroyed at last, their vast body wanting such large quantity of nourishment. When this monster is hungry, it is said to void through the mouth some matter of a sweet scent, which perfumes the whole sea; and by this means it allures and draws all sorts of fishes and animals, even the whales to it, who in whole droves flock thither, and run into the wide opened swallow of this hideous monster, as into a whirlpool, till its belly be well freighted with a copious load of all sorts of fishes and animals, and then it shuts the swallow, and has for the whole year enough to digest and live upon; for it is said to make but one large meal a year. This, though a very silly and absurd tale, is nevertheless matched by another story, every whit as ridiculous, told by my own countrymen, fishermen in the Northern part of Norway. They tell you, that a great ghastly sea monster now and then appears in the main sea, which they call Kracken, and is no doubt the same that the islanders call Hafgufa, of which we have spoken above. They say, that its body reaches several miles in length; and that it is most seen in a calm; when it comes out of the water, it seems to cover the whole surface of the sea, having many heads and a number of claws, with which it seizes all that comes in its way, as fishing boats with men and all, fishes and animals, and lets nothing escape; all which it draws down to the bottom of the sea. Moreover they tell you that all sorts of fishes flock together upon it, as upon a bank of the sea, and that many fishing boats come thither to catch fish, not suspecting that they lie upon such a dreadful monster, which they at last understand by the intangling of their hooks and angles in its body; which the monster feeling, rises softly from the bottom to the surface, and seizes them all; if in time they do not perceive him and prevent their destruction, which they may easily do, only calling it by its name, which it no sooner hears, but it sinks down again as softly as it did rise. They tell you of another sea spectre, which they call the Draw, who keeps to no constant shape or figure, but now appears in one, now in another. It appears and is heard before any misfortunes, as shipwrecks and the like, happen at sea, which it forebodes with a most frightful and ghastly howling; and they say it sometimes utters words like a man. It most commonly diverts itself, in putting all things out of order, after the fishermen are gone at night to rest; and then he leaves behind him a nasty stench. The fishermen will not suffer the truth of this tale to be questioned, but pretend it is confessed. But the most superstitious among them go yet a step farther, and will make you believe, that there appears to them another kind of sea phantom, in the shape of a child in swadling clothes, which they call Marmel, and sometimes draw him out of the sea with their angling hook, when he speaks to them with a human voice. They carry him to their home, and at night they put him into one of their boots, there to rest. In the morning, when they go a fishing again, they take him along with them in their boats, and before they let him go, they set him a task to inform them of all they want to know, upon which they dismiss him.

colony in 64°. This monster was of so huge a size, that coming out of the water, its head reached as high as the mast-head; its body was as bulky as the ship, and three or four times as long. It had a long pointed snout, and spouted like a whale fish; great broad paws, and the body seemed covered with shell work, its skin very rugged and uneven. The under part of its body was shaped like an enormous huge serpent, and when it dived again under water, it plunged backwards into the sea, and so raised its tail aloft, which seemed a whole ship's length distant from the bulkiest part of the body.

OF OTHER FISHES

Of fishes properly so called, the Greenland sea has abundance and of great diversity, of which the largest is called Hay, whose flesh is much like that of the halibut, and is cured in the same manner; being cut into long slices, and hung up to be dried in the sun and in the air, as they cure them in the Northern parts of Norway; but the Greenlanders do not much care for it; its flesh being of a much coarser grain than that of the halibut. This fish has two fins on the back, and six under the belly; the two foremost are the longest, and have the shape of a tongue: the other two middlemost are somewhat broader than the rest, and the hindermost couple near the tail are alike broad before and behind, but shorter than the middlemost: his tail resembles that of the sword fish. There are no bones in him, but gristles only. He has a long snout, under which the mouth is placed like that of the sword fish: he has three rows of sharp pointed teeth; his skin is hard and prickly, of a greyish hue; his length is two or three fathom; he has a great liver, of which they make train oil, the biggest of which makes two or three lasts. It is a fish of prey, bites large pieces out of the whale's body, and is very greedy after man's flesh: he cannot be caught with lines made of hemp, for with his sharp teeth he snaps it off; but with iron chains. And the larger sort are taken with harpoons, as we do the whales. The rest of fishes that haunt the Greenland seas are the halibut, torbut, codfish, haddock, scate, small salmon, or sea-trout of different kinds and sizes (the large salmon not being so frequent in Greenland); and these are very fat and good; they are found in all inlets, and mouths of rivers. Cat-fish is the most common food of Greenlanders, insomuch, that when all other things fail, the cat-fish must hold out, of which there are abundance, both winter and summer. In the spring, towards the month of April, they catch a sort of fish called rogncals, or stone biter; and in May another fish, called lyds or stints: both sorts are very savoury; they frequent the bays and inlets in great shoals. There are also whitings in abundance; but herrings are not to be seen. Moreover there is a kind of fish, which neither myself nor any of my company had ever seen before: this fish is not unlike a bream, only it is prickly with sharp points all over, with a small tail. There are different sizes: the Greenlanders say they are well tasted.

Among the testaceous animals in Greenland the chief are the muscles, of which there are great quantities; they are large and delicate. In some waters I have found of those larger sorts, in which the Norwegians find pearls. These have also pearls, but very small ones, not bigger than the head of a pin. I shall say nothing of the other sea insects, as crabs, shrimps, &c. though they be not rare here; yet lobsters, crawfish, and oysters, I never met with. According to information had of Greenlanders, on the Southern coasts they sometimes catch tortoises in their nets; for they tell you, that they are covered with a thick shell, have claws and a short tail; and moreover that they find eggs in them, like birds' eggs.

OF GREENLAND SEA BIRDS

Amongst the sea fowls the principal are those they call eider-fowl, and ducks; of which there are such numbers, that sometimes sailing along, you find the whole sea covered with them; and when they take their flight, you would think there was no end of them, especially in winter time, when in large flocks, to the number of many thousands, they hover about our colony, morning and evening; in the evening standing in for the bay, and in the morning turning out to sea again. They fly so near the shore, that you may from thence shoot them at pleasure. In the spring they retire towards the sea; for upon the island that lies adjacent to the coast they lay their eggs, and hatch their young ones, which arrive in June and July.

The natives watch them in this season to rob them of their eggs and their young ones. The fine down feathers, which is the best part of this bird, so much valued by others, the natives make nothing of, leaving them in the nests.

There are three sorts of ducks. The first have a broad bill, like our tame duck, with a fine speckled plumage. These build their nests upon the islands as the eider fowls do. The second sort is of a lesser size, their bills long and pointed; they keep most in the bays and in fresh waters, where they nest among the reeds. The third sort are called wood ducks, resembling very much those of the first sort, though somewhat larger in size; the breast is black, the rest of the body grey. These do not propagate in the common way of generation by coupling like other birds, but (which is very surprising) from a slimy matter in the sea, which adheres to old pieces of wood driving in the sea, of which first is generated a kind of muscles, and again in these is bred a little worm, which in length of time is formed into a bird, that comes out of the muscle shell, as other birds come out of egg shells.[30] Besides these there is another sea bird, which the

---

[30] What so many authors of great note relate of the wood ducks, and affirm to be an unquestionable truth, is by as many learned writers treated as an old woman's tale, pretending that such an heterogeneal generation passes the ordinary bounds of nature.

Others (in consideration of so many authors of credit, who affirm that they have been eye witnesses to this strange and wonderful generation) have taken great pains to demonstrate the causes and probability of it physically and philosophically, amongst whom is the learned father Kirkerus, in his *Mundus Subterraneus*; where he maintains, that the semen of this extraordinary generation is neither contained in those old pieces of wood, that drive in the sea, nor in the muscles originally; for a piece of wood cannot produce a living animal, this exceeding the virtue nature has endowed it with; much less the summer froth of the sea, which adheres to the rotten piece of wood, and may produce shells or muscles. Then he forms the question, from whence comes this semen or seed, which produces such a strange fruit as a living bird? which question he strives thus to resolve; that, whereas he has been informed by certain Dutchmen's journals or voyages into the Northern seas, that this sort of birds, peculiar to that climate, make their nest and lay their eggs upon the ice; when the ice by the heat of the sun thaws and breaks asunder, this innumerable quantity of eggs are likewise mashed and crushed to pieces and beaten about by the waves; and that if that part of the egg, which contains the seed, encounters any subject matter proper to foment and brood it, and is received in it loco nutricis, assisted by the temperament of the air, the earth, or the sea, it becomes in due time a perfect bird. This is the renowned father Kirkerus's notion concerning the generation of these birds. But if one examines his reasoning, it is found altogether incoherent: for it was never known, that sea fowls lay their eggs upon the

Norway men call alkes, which in the winter season contribute much to the maintenance of the Greenlanders. Sometimes there are such numbers of them, that they drive them in large flocks to the shore, where they catch them with their hands. They are not so large as a duck, nor is their flesh so well tasted, being more trainy, or oily. The lesser sort of alkes, which also abound here, are more eatable than the large ones. Besides this vast number of sea fowls, there is yet one of a smaller size, by the natives called tungoviarseck, which, for the sake of its beautiful feathers, ought not to be forgot: it has the size and shape of a lark.

Wild geese or grey geese keep to the Northward of Greenland; they are of shape like other geese, somewhat smaller, with grey feathers. They take their flight from other Southern climates over to Greenland every spring, to breed their young ones; which, when grown and able to fly, they carry

---

naked ice, but commonly upon the islands and rocks in the sea, which are surrounded and sometimes covered with ice; and consequently when the ice breaks, and drives away from the islands, the eggs remain still in their nest, without receiving any hurt. And thus the Dutch found it at Nova Zembla, in the year 1569; but what they saw was not the right sort of wood ducks, but what they in Norway call gield ducks; for wood ducks never are seen to couple, nor to lay or hatch their eggs. Secondly, it seems no less absurd to maintain, that eggs, after they are mashed in pieces, and beaten about by the waves, retain as much seminal virtue as will serve to procreate a bird. From whence I infer, that either the information the good father had got from the Dutch voyages was intirely groundless, or this pretended generation goes beyond the bounds of nature. As to the first inference, it is not impossible that the authors who relate this story may have been imposed upon by a common though false report of vulgar and ignorant people; as any one may, that takes a thing for granted upon a bare hearsay, without the attestation of eye witnesses in such a matter. For my part I do not doubt at all of this wonderful generation; for though I have not beheld it with my own eyes, yet I have met with many honest and reasonable men in my native country, who have assured me, that they have found pieces of old, rotten, driven wood in the sea, upon which there hang muscles, in some of which they saw young birds, some half formed, others in full perfection and shape. From whence I conclude, that those fowls spring from no other seed than some clammy and viscous matter floating in the sea, precipitated upon pieces of old rotten wood as aforesaid; of which there is first formed a muscle, and then a little worm in the muscle shell; from whence at last a bird proceeds. And although this may seem to exceed the ordinary bounds set by nature in the procreation of other birds, yet it is observed and confessed, that the sea produces many strange and surprising things, and even living animals, which we cannot affirm to have had being from the first creation; but that by virtue of the primitive blessing God gave the sea to produce, it may yet bring forth many uncommon and wonderful things; as for example, many sorts of sea insects, viz. crabs and the like. And thus the sea or water in general may with reason be stiled *pater et mater rerum*; i. e. "the common parent of things." Nature seems to delight sometimes in forming out-of-the-way things: thus we see divers insects formed out of the very dung of animals; some of which insects often change their kind and shape, *viz.* from a small worm into a flying animal; as flies, beetles, butterflies, and so forth.

along with them and return to the more Southern and milder climates, where they pass the winter season.

In short, I have myself found in Greenland all the several sorts of sea fowls which we have in Norway; as all kinds of mews large and small, which build their nests in the clifts of the highest rocks, beyond the reach of any one; and some upon the little islands, as the bird called terne and the like; whose eggs they gather in great abundance among the stones: the lundes, or Greenland parrot, so called on account of its beautiful plumage and broad speckled bill: the lumbs, the sea-emms, a fowl of a large size, and very small wings, for which reason he cannot fly: besides snipes, and a great number of others; some too common to be enumerated and described here, and others, of which I know not the name.

Chapter 7

TREATS OF THE ORDINARY OCCUPATIONS, AS HUNTING AND FISHING: OF THE TOOLS AND INSTRUMENTS NECESSARY FOR THESE EMPLOYMENTS: OF THE HOUSE IMPLEMENTS AND UTENSILS, &C., OF THE GREENLANDERS

As every nation has its peculiar way of living and of getting their livelihood, suiting their genius and temper to the nature and produce of the country they inhabit; so the Greenlanders likewise have theirs, peculiar to themselves and their country. And though their way and customs may seem to others mean and silly, yet they are such as very well serve their turn, and which we can find no fault with. Their ordinary employments are fishing and hunting: on shore they hunt the rein deer, and at sea they pursue the whales, morses, seals, and other sea animals, as also sea fowls and fishes. The manner of hunting the rein deer has been treated of above in the fifth chapter; but there we took no notice of their bows and arrows, which they make use of in the killing those deer. Their bow is of an ordinary make, commonly made of fir tree, which in Norway is called tenal, and on the back strengthened with strings made of sinews of animals, twisted like thread: the bow string is made of a good strong strap of seal skin, or of several sinews twisted together; the bow is a good fathom long. The head of the arrow is armed with iron, or a sharp pointed bone, with one or more hooks, that it may keep hold, when shot into a deer's body. The arrows they shoot birds with are at the head covered with one or more pieces of bone blunt at the end, that they may kill the fowl without tearing the flesh. The sea fowls are not shot with arrows, but with darts, headed with bones or iron, which they throw very dexterously, and with so steady a hand at a great distance, that nobody can hit surer with a gun. They are more frequently employed at sea than on shore; and I confess they surpass therein most other nations; for their way of taking whales, seals, and other sea animals is by far the most skilful and most easy and handy.

When they go whale catching, they put on their best gear or apparel, as if they were going to a wedding feast, fancying that if they did not come cleanly and neatly dressed, the whale, who cannot bear slovenly and dirty habits, would shun them and fly from them. This is the manner of their expedition: about fifty persons, men and women, set out together in one of the large boats, called kone boat; the women carry along with them their sewing tackles, consisting of needles and thread, to sew and mend their husbands' spring coats, or jackets, if they should be torn or pierced through, as also to mend the boat, in case it should receive any damage; the men go in search of the whale, and when they have found him they strike him with their harpoons, to which are fastened lines or straps two or three fathoms long, made of seal skin, at the end of which they tie a bag of a whole seal skin, filled with air, like a bladder; to the end that the whale, when he finds himself wounded, and runs away with the harpoon, may the sooner be tired, the air bag hindering him from keeping long under water. When he grows tired and loses strength, they attack him again with their spears and lances, till he is killed, and then they put on their spring coats, made of dressed seal skin, all of one piece, with boots, gloves, and caps, sewed and laced so tight together that no water can penetrate them. In this garb they jump into the sea, and begin to slice the fat of him all round the body, even under the water; for in these coats they cannot sink, as they are always full of air; so that they can, like the seal, stand upright in the sea: nay they are sometimes so daring, that they will get upon the whale's back while there is yet life in him, to make an end of him and cut away his fat.

They go much the same way to work in killing of seals, except that the harpoon is lesser, to which is fastened a line of seal skin six or seven fathoms long, at the end of which is a bladder or bag made of a small seal skin filled with air to keep the seal, when he is wounded, from diving under the water, and being lost again. In the Northern parts, where the sea is all frozen over in the winter, they use other means in catching of seals. They first look out for holes, which the seals themselves make with their claws, about the bigness of a halfpenny, that they may fetch their breath; after they have found any hole, they seat themselves near it upon a chair made for this purpose; and as soon as they perceive the seal come up to the hole and put his snout into it for some air, they immediately strike him with a small harpoon, which they have ready in their hand, to which harpoon is fastened a strap a fathom long, which they hold with the other hand. After he is struck, and cannot escape, they cut the hole so large, that they may get him up through it; and as soon as they have got his head above the ice, they can kill him with one blow of the fist.

A third way of catching seals is this: they make a great hole in the ice, or, in the spring, they find out holes made by the seals, through which they get upon the ice to lie and bask themselves in the sun. Near to these holes they place a low bench, upon which they lie down upon their belly, having first made a small hole near the large one, through which they let softly down a perch, sixteen or twenty yards long, headed with a harpoon, a strap being fastened to it, which one holds in his hand, while another (for there must be two employed in this sort of capture) who lies upon the bench with his face downwards, watches the coming of the seal, which when he perceives, he cries "Kæ;" whereupon he, who holds the pole, pushes and strikes the seal.

The fourth way is this: in the spring, when the seals lie upon the ice near holes, which they themselves make to get up and down, the Greenlanders, clothed with seal skins, and a long perch in their hand, creep along upon the ice, moving their head forwards and backwards, and snorting like a seal, till they come so near him, that they can reach him with the perch and strike him. A fifth manner of catching seals is, when in the spring the current makes large holes in the ice, the seals flock thither in great shoals; there the natives watch their opportunity to strike them with their harpoons, and haul them upon the ice. There is yet a sixth way of catching seals, when the ice is not covered with snow, but clear and transparent; then the catchers lay under their feet foxes or dogs' tails, or a piece of a bear's hide, to stand upon and watch the animal, and when by his blowing and snorting they find what course he takes, they softly follow him and strike him.

In fishing they make use of hooks and angles of iron or bones. Their lines are made of whalebones cut very small and thin, and at the end tacked together; and with such lines they will draw one hundred fishes to one which our people can catch with their hemp lines. But to catch halibut they use strong lines made of seal skin, or thick hemp lines.

Their way of fishing the small salmon or sea trout is this: at low water they build small enclosures with stone, near the river's mouth, or any other place where the salmon runs along; and when it begins to flow, and the tide comes in, the salmon retreats to the river, and in high water passes over the enclosure, and remains in the river till the water again falls; then the salmon wants to go to sea again; but the fishermen way-lay him at the enclosure and stop his passage. And soon after, when the water is quite fallen and it is low ebb, the salmon remains upon dry land, and may be caught with hands. And where they are left in holes, they take them with an instrument made for this purpose, viz. a perch headed with two sharp hooked bones, or with one or two iron hooks.

The rogn fish, or roe fish, so named from the great quantity of roe that is found in it, as he is commonly found in shallow water and upon the sands, so he is caught like the salmon with the before-mentioned instrument. There is such abundance of these fishes, that, as they cannot consume them all fresh, they are obliged to dry them on the rocks, and keep them for winter provision. When roe fish catching is over, which happens in the month of May, then the Greenlanders retire into the bays and creeks, where the lod or stint fishing then takes place. There are such numberless shoals of them near the shore, that they catch them in a kind of sieves fastened upon long poles, and throw them upon the shore; they open and dry them upon the rocks, keeping them for their winter stock. This fish is not agreeable, nor reckoned wholesome, when eaten fresh; besides they have a nauseous smell, but when dried they may pass. The natives eat them with a bit of fat, or soused in train oil: and so of all other sorts of fishes, what the Greenlanders cannot consume fresh they dry upon the rocks in the sun, or in the wind, and lay them up for the winter.

Now as to the Greenland boats, there are two sorts of them; the one of which the men alone make use, is a small vessel sharp and pointed at both ends, three fathoms in length, and at most but three quarters of a yard broad with a round hole in the midst, just large enough for a man's body to enter it, and sit down in it, the inside of the boat is made of thin rafts tacked together with the sinews of animals, and the outside is covered with seal skins, dressed and without hair; no more than one can sit in it, who fastens it so tight about his waist, that no water can penetrate it. In these small boats they go to sea, managing them with one oar of a fathom in length, broad at both ends, with which they paddle sometimes on one side, and sometimes on the other, with so much swiftness, that they are said to row ten or twelve Norway miles in a day. They chiefly make use of them in catching of seals and sea fowls, which they can approach on a sudden and unawares; whereas we in our large boats can very seldom come so near as to touch them. They do not fear venturing out to sea in them in the greatest storms, because they swim as light upon the largest waves as a bird can fly; and when the waves come upon them with all their fury, they only turn the side of the boat towards them to let them pass, without the least danger of being sunk: though they may happen to be overset, yet they easily raise themselves again with their paddle; but if they are overset unawares (as it often happens) and the boat be not close and tight about their waist, they are inevitably drowned.

The other kind of boats are large and open, like our boats, some of them twenty yards long; and these are called kone boats, that is, women's boats, because the women commonly row them; for they think it unbecoming a man to row such a boat, unless great necessity requires it: and when they first set out for the whale fishing, the men sit in a very negligent posture, with their faces turned towards the prow, pulling with their little ordinary paddle; but the women sit in the ordinary way, with their faces towards the stern, rowing with long oars. The inside of these boats is composed of thin rafts, and the outside clothed with thick seal skins. In these boats they transport their baggage, as tents and the like household furniture, when they go to settle in some distant places in quest of provision. In these boats they also carry sails, made of the bowels and entrails of seals. The mast is placed foremost on the prow, and as the sail is broad at the upper end, where it is fastened to the yard and narrow at the lower end, so they neither want braces nor bowlines and sheet ropes, and with these sails they sail well enough with the wind, not otherwise. These boats, as they are flat-bottomed, can soon be overset.

The men meddle with no work at home but what concerns their tools for hunting and fishing tacklings, *viz.* their boats, bows, arrows, and the like. All other work, even of building and repairing their houses, belongs to the women. As dexterous and skilful as the men are at their work, so the women are not behindhand with them, but according to their way and manner deserve to be praised and admired.

Chapter 8

OF THE INHABITANTS, THEIR HOUSES, AND HOUSE FURNITURE

It is undoubted, that the modern inhabitants of Greenland are the offspring of the Schrellings, especially those that live on the Western coast;

and there may be some mixture, for aught we know, of the ancient Norway colonies that formerly dwelled in the country, who in length of time were blended and naturalized among the natives, which is made probable by several Norway words found in their language. For, although the Norway colonies were destroyed, yet there were, no doubt, some remains of them, which joined with the natives and became all one nation. With these inhabitants all the sea coasts are peopled, some more and some less.

The coast is pretty populous in the Southern parts, and on the North in 68° and 69°; though, compared to other countries, it is in the main but thinly inhabited. In the inner parts of the country nobody lives, except at certain times in the summer season, when they go rein deer hunting. The reason of this is, that (as has been said above) the whole upland country is perpetually covered with ice and snow.

As to their houses or dwelling places, they have one for the winter season and another for the summer. Their winter habitation is a low hut built with stone and turf, two or three yards high, with a flat roof. In this hut the windows are on one side, made of the bowels of seals dressed and sewed together, or of the maws of halibut, and are white and transparent. On the other side their beds are placed, which consist in shelves or benches made up of deal boards raised half a yard from the ground; their bedding is made of seal and rein deer skins.

Several families live together in one of these houses or huts; each family occupying a room by itself, separated from the rest by a wooden post, by which also the roof is supported; before which there is a hearth or fireplace, in which is placed a great lamp in the form of a half moon seated on a trevet; over this are hung their kettles of brass, copper, or marble, in which they boil their victuals: under the roof, just above the lamp, they have a sort of rack or shelf, to put their wet clothes upon to dry. The fore door or entry of the house is very low, so that they must stoop, and most creep in upon all fours, to get in at it; which is so contrived to keep the cold air out as much as possible. The inside of the houses is covered or lined with old skins, which before have served for the covering of their boats. Some of these houses are so large, that they can harbour seven or eight families.

Upon the benches or shelves, where their beds are placed, is the ordinary seat of the women, attending their work of sewing and making up the clothing. The men with their sons occupy the foremost parts of the benches, turning their back to the women: on the opposite side, under the windows, the men belonging to the family, or strangers, take their seats upon the benches there placed.

I cannot forbear taking notice, that though in one of these houses there be ten or twenty train lamps, one does not perceive the steam or smoke thereof to fill these small cottages: the reason, I imagine, is, the care they take in trimming those lamps, *viz.* they take dry moss, rubbed very small, which they lay on one side of the lamp, which, being lighted, burns softly and does not cause any smoke, if they do not lay it on too thick, or in lumps. This fire gives such a heat, that it not only serves to boil their victuals, but also heats the room to that degree, that it is as hot as a bagnio. But for those who are not used to this way of firing, the smell is very disagreeable, as well by the number of burning lamps, all fed with train oil, as on account of divers sorts of raw meat, fishes, and fat, which they heap up in their habitations; but especially their urine tubs smell most insufferably, and strike one, that is not accustomed to it, to the very heart.

These winter habitations they begin to dwell in immediately after Michaelmas, and leave them again at the approach of the spring, which commonly is at the latter end of March; and then for the summer season lodge in tents, which are their summer habitations. These tents are made of rafts or long poles, set in a circular form, bending at the top, and resembling a sugar loaf, and covered with a double cover, of which the innermost is of seal or rein deer skins with the hairy side inward (if they be rich), and the outermost also of the same sort of skins, without hair, dressed with fat, that the rain may not pierce them. In these tents they have their beds, and lamps to dress their meat with; also a curtain made of the guts or bowels of seals sewed together, through which they receive the day light instead of windows. Every master of a family has got such a tent, and a great woman's boat, to transport their tents and luggage from place to place, where their business calls them.

Chapter 9

THE GREENLANDERS' PERSONS, COMPLEXION, AND TEMPERAMENT

The Greenlanders, as well man as womankind, are well shaped and proportioned, rather short than tall, and strong built, inclined to be fat and corpulent; their faces broad, thick lips, and flat nosed; their hair and eyes black, their complexion a very dark tawny; though I have seen some pretty fair. Their bodies are of a vigorous constitution. There are seldom found any sick or lame, and but few distempers are known among them, besides weakness of the eye-sight, which is caused by the sharp and piercing spring winds, as well as the snow and ice, that hurt the sight.

I have met with some that seemed infected with a kind of leprosy; yet (what is surprising to me), though they converse with others, and lay with them in one bed, it is not catching. They that dwell in the most Northern parts are often miserably plagued with dysenteries or bloody fluxes, breast diseases, boils, and epilepsy, or falling sickness, &c. There were no epidemical or contagious diseases known among them, as plague, small-pox, and such like, till the year 1734, when one of the natives, who with several others were brought over to Denmark, and together with his companions had the small-pox at Copenhagen, coming home again to his native country brought the infection amongst them; of which there were swept away in and about the colony about two thousand persons. For as the natives as well as the animals of this climate are of a hot nature, they cannot bear the outward heat, much less the inward, caused by this burning distemper, which inflames the mass of blood to that degree, that it cannot, by any means, be quenched. They are very full of blood, which is observed by their frequent bleeding at the nose.

Few of them exceed the age of fifty or sixty years; many die in the prime of their life, and most part in their tender infancy; which is not to be wondered at, considering they are quite destitute of all sorts of medicines, and ignorant of all that may strengthen and comfort sick bodies. To supply which defects, they know of nothing better than to send for their divines, which they name angekuts, who mutter certain spells over the sick, by which they hope to recover.

For outward hurts, as wounds, cuts of knives, and the like, they sew or stitch them together. If any grow blind, as it often happens to them, the eye being covered over with a white skin, they make a small hook with a needle, which they fasten into this skin, to loosen it from the eye, and then with a knife they pull it off. When children are plagued with worms, the mother puts her tongue (salva vericâ) into the anus of the children, to kill them. Burnt moss with train oil mixed together serves for plaisters to fresh wounds; or they cover them with a piece of the innermost rind of a tree, and it will heal of itself.

The Greenlanders are commonly of a phlegmatic temper, which is the cause of a cold nature and stupidity: they seldom fly into a passion, or are much affected or taken with anything, but of an insensible, indolent mind. Yet I am of opinion, that what contributes most to this coldness and stupidity is want of education and proper means to cultivate their minds. In which opinion I am confirmed by the experience of some who had for some time conversed with us, especially the young ones, who easily have taken all that they have seen or heard among us, whether it was good or bad. I have found some of them witty enough, and of good capacity.

Chapter 10

THE CUSTOMS, VIRTUES, AND VICES, AND THE MANNERS OR WAY OF LIFE OF THE GREENLANDERS

Though the Greenlanders are as yet subject to no government, nor know of any magistrates, or laws, or any sort of discipline; yet they are so far from being lawless or disorderly, that they are a law to themselves; their even temper and good nature making them observe a regular and orderly behaviour towards one another. One cannot enough admire how peaceably, lovingly, and united they live together; hatred and envy, strifes and jars are never heard of among them.[31] And although it may happen that one bears a grudge to another, yet it never breaks out into any scolding or fighting; neither have they any words to express such passions, or any injurious and provoking terms of quarrelling. It has happened once or twice, that a very wicked and malicious fellow, out of a secret grudge, has killed another; which none of the neighbours have taken notice of, but all let it pass with a surprising indolence; save the next kindred to the dead, if he finds himself strong enough, revenges his relation's death upon the

---

[31] When they see our drunken sailors quarrelling and fighting together, they say we are inhuman; that those fighters do not look upon one another to be of the same kind. Likewise, if an officer beats any of the men, they say, such officer treats his fellow creatures like dogs.

murderer. They know of no other punishment; but those old women called witches, and such as pretend to kill or hurt by their conjuring; to such they show great rigour, making nothing of killing and destroying them without mercy. And they pretend that it is very well done; those people not deserving to live, who by secret arts can hurt and make away with others.

They have as great an abhorrence of stealing or thieving among themselves, as any nation upon Earth; wherefore they keep nothing shut up under lock and key, but leave everything unlocked that everybody can come at it, without fear of losing it.

This vice is so much detested by them, that if a maiden should steal anything, she would thereby forfeit a good match. Yet if they can lay hands upon anything belonging to us foreigners, they make no great scruple of conscience about it. But, as we now have lived sometime in the country amongst them, and are looked upon as true inhabitants of the land, they at last have forbore to molest us anymore that way.

As to the transgression of the seventh commandment, we never have found them guilty in that point, either in words or deeds, except what passes amongst the married people in their public diversions, as we shall see hereafter.

As for what we call civility and compliments, they do not much trouble themselves about them; they go and come, meet and pass one another, without making use of any greeting or salutation: yet they are far from being unmannerly or uncivil in their conversation; for they make a difference among persons, and give more honour to one than to another, according to their merit and deserts. They never enter any house where they are strangers, unless they are invited, and when they come in, the master of the house, to whom they pay the visit, shows them the place where they are to take their seat.

As soon as a visitor enters the house, he is desired forthwith to strip naked, and to sit down in this guise like all the rest; for this is the grand fashion with them to dry the clothes of their guest. When victuals are put before him, he takes care not to begin eating immediately, for fear of being looked upon as starved, or of passing for a glutton. He must stay till all the family is gone to bed before he can lie down, for to them it seems unbecoming that the guest goes to rest before the landlord. Whenever a stranger comes into a house, he never asks for victuals, though never so hungry; nor is there any need he should; for they generally exercise great hospitality, and are very free with what they have; and what is highly to be admired and praiseworthy, they have most things in common; and if there be any among them (as it will happen) who cannot work or get his livelihood, they do not let him starve, but admit him freely to their table, in which they confound us Christians, who suffer so many poor and distressed mortals to perish for want of victuals.

Finally, the Greenlanders, as to their manners and common way of life, are very slovenly, nasty, and filthy; they seldom wash themselves,[32] will eat out of plates and bowls after their dogs, without cleansing them; and (what is most nauseous to behold), eat lice and such like vermin, which they find upon themselves or others. Thus they make good the old proverb, what drips from the nose falls into the mouth, that nothing may be lost. They will scrape the sweat from off their faces with a knife, and lick it up. They do not blush to sit down and ease themselves in the presence of others. Every family has a urine tub placed before the entry, in which they make water, and leave it so standing till it smells most insufferably, for they put in it the skins, which are to be dressed, to soak or steep, which affords not the most agreeable scent; to the encreasing of which the rotten pieces of flesh meat and fat thrown under their benches contributes a great deal; so that delicate noses do not find their account among them. Yet through long custom the most nauseous things become more supportable.

Notwithstanding, however, their nasty and most beastly way of living, they are very good natured and friendly in conversation. They can be merry and bear a joke, provided it be within due bounds. Never any of them has offered in the least manner to hurt or to do harm to any of our

---

[32] The way the men wash themselves is to lick their fingers (as the cat does his paws) and rub their eyes with them to get the salt off, which the sea throws into their face. The women wash themselves in their urine, that their hair may grow, and to give it (according to their fancy) a fine smell. When a maiden has thus washed herself, their common saying is *niviarsiarsuanerks*, that is, she smells like a virgin maid. Thus washed they go into the cold air, and let it freeze, which shows the strength of their heads, and it well becomes foreigners to do so.

people, unless provoked to it. They fear and respect us as a nation far superior to theirs in valour and strength.

Chapter 11

OF THEIR HABITS, AND WAY OF DRESSING

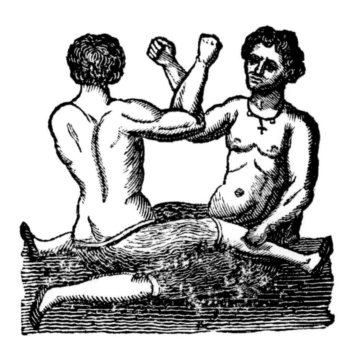

Their clothes are, for the most part, made of reindeer and seal skin, as also of bird's skin nicely dressed and prepared. The men's habits are a coat or jacket, with a cap or hood sewed to it, to cover the head and shoulders, in the fashion of a domino, or monk's hood. This coat reaches down to the knees. Their breeches are very small, not coming above their loins, that they may not hinder them in getting into their small boats. And as they wear no linen, the hair of the skins the coat is made of is turned inward to keep them warm. Over this coat they put on a large frock, made of seal skin dressed and tanned, without hair, in order to keep the water out; and thus they are dressed when they go to sea.

Between the leathern frock and the under coat they wear a linen shirt, or, for want of linen, made of seal's guts; which also helps to keep out the water from the under coat. Of late they appear sometimes in more gaudy dresses, as shirts made of striped linen, and coats and breeches of red and blue stuffs, or cloth, which they buy of ours, or the Dutch merchants, but fashioned after their own way; in these they make parade and feast, when they keep holidays on shore. The stockings they wore formerly were made of rein deer, or seal's skin, but now they like better our sort of worsted stockings, of different colours, white, blue, and red, which they buy of us. Their shoes and boots are made of seal's skins, red or yellow, well dressed and tanned; they are nicely wrought, with folds behind and before, without heels, and fit well upon the foot.[33]

The only difference between the dress of the men and the women is, that the women's coats are higher on the shoulders and wider than the men's, with higher and larger hoods. The married women, that have got children, wear much larger coats than the rest, most like gowns, because they must carry their children in them upon their backs, having got no other cradle or swadling clothes for them. They wear drawers, which reach to the middle of the thigh, and over them breeches: the drawers they always keep on, and sleep in them. Their breeches come down to the knee: these they do not wear in the summer, nor in the winter, but when they go abroad; and as soon as they come home they pull them off again. Next to their body they wear a waistcoat made of young fawns' skins, with the hairy side inward. The coat, or upper garment, is also made of fine coloured swans' skins (or, in defect of that, of seal skins) trimmed and

---

[33] In the summer they wear short frocks, as also in winter, when they work on the ice in the bays; but then they put a white covering over it, that they may not frighten the seals.

edged with white, and nicely wrought in the seams, and about the brim, which looks very well. Their shoes and boots, with little difference, are like those of the men. Their hair, which is very long and thick, is braided and tied up in a knot, which becomes them well. They commonly go bareheaded, as well without as within doors; nor are they covered with hoods, but in case it rains or snows. Their chief ornament and finery is to wear glass beads of divers colours, or corals about the neck and arms, and pendants in their ears. They also wear bracelets, made of black skin, set with pearls, with which they also trim their clothes and shoes.

The Greenland sex have, besides this, another sort of embellishment, viz. they make long black strokes between the eyes on the forehead, upon the chin, arms, and hands, and even upon the thighs and legs: these they make with a needle and thread made black. And though this to others seems a wrong way of embellishing, yet they think it very handsome and ornamental. And they say that those who do not thus deform their faces, their heads shall be turned into train tubs, which are placed under the lamps in Heaven, or the land of souls.

They keep their clothes pretty clean, though in other things, especially in their victuals, they are not so nice, chiefly the women, who have got children, are very dirty and slovenly, well knowing, that they cannot be repudiated, or sent a packing. But those wretches that are barren, or whose children are dead, and do not know the moment they may be sent away, are obliged to take more care of their cleanness and property, that they may please their husbands.

Chapter 12

OF THEIR DIET, AND MANNER
OF DRESSING THEIR VICTUALS

The Greenlanders' provision and victuals are flesh and fish meat (for the country affords no other kind of provision) as rein deer, whales, seals, hares, and rypes, or white partridges, and all sorts of sea fowls. They eat their flesh meat sometimes raw, sometimes boiled, or dried in the sun or wind; but their fish meat is always thoroughly done, or they eat it dried in the sun or air, as salmon, roe-fish, halibut, or the small stints, which, in the months of May and June, they catch in great abundance, and keep them cured and dried for winter provisions. And whereas, in the winter season, it is very rare to get seals, except in the most Northern parts where they take them upon the ice; so they make all the provision of them they can get in the fall, and bury them under the snow, until the winter comes on, when they dig them up, and eat them raw and frozen as they are. Their drink is nothing but water, and not, as some writers have wrongly pretended, train oil; for they do not so much as eat the fat, but only in sauces to their dried fish.

Furthermore, they put great lumps of ice and snow into the water they drink, to make it the cooler to quench their thirst. They are, taking them in general, very hoggish and dirty in their eating and dressing of their victuals; they never wash, cleanse, or scour the kettles, pots, or dishes, in which they dress, and out of which they eat their victuals; which when dressed, they often lay down upon the dirty ground, which they walk upon, instead of tables. They will, with so great an appetite and greediness, feed upon the rotten and stinking seal flesh, that it turns the stomach of any hungry man who looks upon them. They have no set time for their meals, every man eats when he is hungry, except when they go to sea, and then their chief repast is a supper, after they are come home in the evening; and he, whose supper is first ready, calls his neighbours to come and partake of it, as he does again with them reciprocally; and so it goes round from one to another.

The women do not eat in company with the men, but separately by themselves; and in the absence of their husbands, when gone a fishing, they being left to themselves, invite one another, and make grand cheer. And as they eat heartily, when they can come at it, so they can as well endure hunger, when scarcity of provision requires it. It has been observed, that in great scarcity, they can live upon pieces of old skins, upon reets, or sea weeds, and other such trash. But the reason why they can endure hunger better than we foreigners, I take to be, their bodies being so squat and corpulent, their fat yielding them matter of nourishment within themselves, for a while, till it be consumed.

Besides the fore-mentioned provisions, they also eat a sort of reddish sea weed, and a kind of root, which they call tugloronet, both dressed with fat or train oil; the dung of the rein deer, taken out of the guts, when they cleanse them; the entrails of partridges, and the like out-cast, pass for dainties with them. They make likewise pancakes of what they scrape off the inside of seal skins, when they dress them. In the summer they boil their meat with wood, which they gather in the field, and in winter time over their lamps in little kettles of an oval figure, made of brass, copper, or marble, which they make themselves.

To kindle the fire, when extinguished, they make use of this expedient, which shows their ingenuity: they take a short block of dry fir tree, upon which they rub another piece of hard wood, till, by the continued motion, the fir catches fire. When we first came among them, they did not like to taste any of our victuals, but now they are glad to get some of it, especially bread and butter, which they like mightily, but they do not much care for our liquors; yet notwithstanding, some of them, who have lived some time among us, have learnt to drink wine and brandy, and never refuse it, when it is offered them. But as for tobacco, they do not at all like it, nor can they bear the smell or smoke of it.

Chapter 13

OF THEIR MARRIAGES, AND EDUCATION OF THEIR CHILDREN

The most detestable crime of polygamy, which reigns so much among the Heathens, the Greenlanders are not so much addicted to; for commonly they are contented with one wife. There are some, but very few, that keep two, three, or four wives: but these pass for heroes or more than ordinary men, in that, by their industry, they are able to subsist so many wives and children. And what is remarkable, before our arrival, there was never heard of such a thing as jealousy among those wives, but they agreed very well together, though the first wife was reckoned the mistress. Since our arrival, as we have informed them of the word and will of God, importing, that in the beginning the All-wise Creator made one man and one woman, to live in matrimony as husband and wife, there has been some resentment in the wives, when their husbands have had a mind to take any other besides them; they have addressed themselves to me, and desired me to put a stop to such a proceeding. Also when I have instructed them in their catechism and the Christian doctrine, they have always put me in mind, not to forget fully to instruct their husbands in the duties of the seventh commandment.

Some time passed before we could learn how the men behaved themselves with regard to other men's wives, or the women vice versa, till at last we perceived them not to be over scrupulous in this matter, of which we were more fully convinced, by hearing of a certain illegal game used among them; which is this. A number of married men and women meet together at an assembly; where, after they have taken their fill of feasting and revelling, they begin singing and dancing, according to their own way; and in the mean while one after another take a trip with each other's wife, behind a curtain or hangings made of skins at one end of the house, where their beds are placed, and there divert themselves. Those are reputed the best and noblest tempered, who, without any pain or reluctancy, will lend their friends their wives.

But, as I observed above, none but married people frequent these sort of games, which, they imagine, is not unbecoming. Especially the women think themselves happy, if an angekkok, or prophet, will honour them with his caresses: there are even some men so generous, that they will pay the angekkok for it; chiefly if they themselves have no children; for they fancy that an angekkok's child will be more happy and better qualified for business than others.

Maidens, on the contrary, and unmarried women, observe much better the rules of modesty and continency; for I never saw any of them entertain any loose or slippery conversation with young men; or show the least inclination to it either in words or deeds. During fifteen full years that I lived in Greenland, I did not hear of more than two or three young unmarried women, who had been guilty of incontinence; because it is reckoned the greatest of infamies. It is remarkable, that natural decency is observed by them; for they refrain from marrying their next relations, even in the third degree, taking such matches to be unwarrantable and quite unnatural. It is likewise reckoned uncouth and blameable, if a lad and a girl, that have served and been educated in one family, should desire to be married together; for they look upon them as brother and sister.

The ceremonies they use in their marriages and weddings are as follow:—When a young man likes a maiden, he commonly proposes it to their parents and relations on both sides; and after he has obtained their consent, he gets two or more old women to fetch the bride (and if he is a stout fellow, he will fetch her himself). They go to the place where the young woman is, and carry her away by force; for though she ever so much approves of the match, yet out of modesty she must make as if it went against the grain, and as if she was much ruffled at it; else she will be blamed and get an ill name, as if she had been a love-sick wench. After she is brought to the house of the bridegroom, she keeps for some time at a distance, and sits retired in some corner, upon the bench, with her hair dishevelled, and covering her face, being bashful and ashamed. In the mean while the bridegroom uses all the rhetorick he is master of, and spares no caresses to bring her to a compliance with his ardent wishes; and the good girl being at length persuaded and prevailed with, yields kindly to his ravishing embraces; and then they lie down together, and so the wedding is over. But sometimes they take a shorter way to go to work, which is to gratify their inclinations without the advice or consent of the parents.[34] Nevertheless their matrimony is not of so indissoluble a nature but that the husbands often repudiate and put away their wives, if either they do not suit their humours, or else, if they are barren and do not bring forth children (which they hold to be very ignominious), and marry others. But if they have children by them, they bear a great deal with them, and keep them for life. It is not rare to see that a man beats his wife, and gives her black eyes, for her obstinacy and stubbornness sake; however they are soon reconciled and good friends again, without bearing any grudge. For, according to them, it signifies nothing, that a man beats his wife; but they do not like that a master should drub a servant maid. Likewise they think it heinous that a mother chastises her children; and if she falls foul of her maid, it is with them unpardonable; and such a woman gets an ill name.

If one of the party dies, the relict, whether husband or wife, is at liberty to marry again.

The women are of a very hardy and strong nature, which they chiefly show in their child-bearing; for as soon as it is over, they will go to work and do their ordinary business as usual. But sometimes they pay very dear

---

[34] When a man sends for his son's bride, to be conducted to his house, if he be in good circumstances he makes a great feast; and throws out for prizes several presents of poles, rafts, knives, and other toys. The same is practised the day following after the bedding of the new-married couple. If they have children before the year is past, or if they often breed, they are blamed, and compared to dogs. A new married woman is ashamed for having changed her condition for a married state.

for this bravery, it costing them their lives. The day after their delivery they go abroad to work, being girt with a waist belt two or three inches broad, which they also wore before their delivery. As soon as the child is born, the mother dips her finger into water, and rubs the child's lips with it; or she puts a little bit of snow into its mouth, saying, "Imekautit," which signifies, Thou hast drunk a good deal; and when she eats, she takes a bit of fish, and holds it to the child's mouth, and shakes her hand, with this word, "Aiparpotit," that is to say, Thou hast eat and kept me company. They cut the navel-string, not with a knife, but with a muscle shell, or they bite it off with their teeth; and when the string is dry they use it as an amulet.

They hold a chamber pot over the head of the woman in labour, imagining that it helps to hasten her delivery. The child being a year old, the mother slabbers and licks it all over, from head to foot, that it may grow hale and strong. They seldom bear twins, but monsters are often brought forth. In the year 1737 a woman, in the Bay of Disco, was delivered of a hideous monster; the eyes were placed on the side of the nose: it had a pointed snout and no ears. Instead of hands and feet it had paws, and very thick thighs. Its front was covered with hair like those of a rein deer, and the sides were covered with something like a white skin of a fish. In the same place another monstrous birth was seen in the year 1739, without a head, four-footed, with long nails, like claws; it had a mouth upon the breast, and claws upon the back.

They have a very tender love for their children, and the mother always carries her infant child about with her upon her back, wrapped up in her coat wherever she goes, or whatever business she has in hand, for they have no other cradles for them. They suckle them till they are three or four years old or more; because in their tender infancy they cannot digest the strong victuals that the rest must live upon.

The education of their children is what they seem little concerned about; for they never make use of whipping or hard words to correct them, when they do anything amiss, but leave them to their own discretion. Notwithstanding which, when they are grown, they never seem inclined to vice or roguery, which is to be admired. It is true, they show no great respect to their parents in their outward forms, but always are very willing to do what they order them; though sometimes they will bid their parents do it themselves. They are under the care of their parents, boys as well as girls, till they are married; afterwards they shift for themselves, yet so, that they continue to dwell in the same house, or under the same roof with their fathers, together with other kindred and relations; and what they get, they all enjoy in common.

Chapter 14

HOW THE GREENLANDERS MOURN AND BURY THEIR DEAD FRIENDS

When any person dies, they take what belongs to him, as house-furniture, utensils, and clothing, and throw it all out into the field, that by touching of them they may not become unclean, or any misfortune befal them on that account: and all that live in the same house are obliged to carry out anything of their goods that is new and has not been used; but in the evening they bring them all back again, for then they say the stench of the dead body is quite dissipated. Then they begin to lament and mourn for their dead friend, with tears and ghastly howlings, which they continue for an hour, and then the nearest relations take the body and carry it to the grave, made up of stones thrown together in a heap, under which they bury him dressed in his best clothes, and well wrapt up in skins of rein deer or seals, with his legs bent under his back. Near the burying place they lay his utensils, viz. his boat, bows, arrows, and the like; and if it be a woman, her needles, thimbles, and the like; not that they believe they stand in need of those things, when they are come to the land of souls, or in the other world, whither they are retired, but for the aversion they have for those things: lest by refreshing the memory of the deceased, they might renew their grief and sorrow for his loss; for if they should bewail him and weep too much, they think he will endure the more cold where he is.

They think themselves unclean if they touch anything belonging to the deceased; as likewise he that has carried him to the grave, and buried him, is reckoned unclean for some time, and dares not do certain things: nay, not only the kindred and relations of the deceased, but likewise every one that has lived in the same house with him, are obliged to abstain from certain victuals and work, for a while, according to the direction of the angekkuts or divines.

The women never wash themselves during their mourning time, nor appear well dressed, or with braided and tied up hair, but dishevelled, and hanging about the face. They must put on their hood as often as they go out of doors, which is not customary at other times: but they believe they otherwise should soon die.

They bewail their dead long enough: for, as often as any of their friends and acquaintance come from other places to see them, the first thing they do is to sit down in great sadness, and weep and bemoan the loss of their deceased friend: after which they are comforted with good cheer. But if the deceased has left no friend or relation behind him, he may lie long enough where he died, whether at home or abroad before anybody comes and buries him. If a person dies in the house, his body must not be carried through the ordinary entry of it, but conveyed out at the window; and if he dies in a tent, he is brought out at the back part of it. At the funeral a woman lights a stick in the fire, brandishing the same and saying piklerrukpok, that is, Here is no more to be got.

When little children die and are buried, they put the head of a dog near the grave, fancying, that children having no understanding, they cannot by themselves find the way, but the dog must guide them to the land of the souls.

Chapter 15

THEIR PASTIMES AND DIVERSIONS, AS ALSO THEIR POETRY

The Greenlanders have several kinds of sports and recreations, with which they pass their time, when they have nothing else to do, or when they visit one another: of which these are the most remarkable. When they meet together for diversion's sake, the first step made is always banqueting and revelling, where they stuff themselves with all the dainty bits and the best cheer the country affords; as rein deer and seal flesh dried or boiled; and the tail of a whale, which they reckon among the greatest delicacies. Of these things they eat very greedily; for it is a great honour done to the landlord who treats, that his guests, when come home, complain that their belly was too small, and that it was ready to burst.

After the repast, they get up to divert themselves in this manner: one of the company takes a drum, which is made of a broad wooden hoop, or of the rib of a whale, covered with a thin skin, with a handle to it; which drum he beats with a stick, singing at the same time songs, either concerning the common affairs in general, or his own private ones in particular. In which, at the end of each verse, the whole chorus of men and women join with him.

He that can play the most odd and comical gestures, and play the most ridiculous tricks with his face, head, and limb, turning them awry, passes for the most ingenious fellow; as he by his awkward and out of the way postures can make others laugh.

They show their wit chiefly in satirical songs, which they compose against one another; and he, that overcomes his fellow in this way of debate, is admired and applauded by the rest of the assembly. If anybody conceives a jealousy, or bears a grudge to another upon any account, he sends to him, and challenges him to a duel in such or such assembly; where he will fight it out with him in taunting ditties. Whereupon the defied, in defence of his honour, prepares his weapons, and does not fail to appear at the time and place appointed, if his courage do not forsake him. When the assembly is met, and the combatants arrived, everybody being silent and attentive to hear what end the combat will take, the challenger first enters the lists, and begins to sing, accompanying it with the beat of his drum. The challenged rises also, and in silence listens, until his champion or adversary has done singing. Then he likewise enters the lists, armed with the same weapons, and lays about his party the best he can. And thus they alternately sing as long as their stock of ditties lasts. He that first gives over, is reckoned overcome and conquered. In this sort of taunting ditties they reproach and upbraid one another with their failings. And this is their common way of taking vengeance.

There is not to be expected great ingenuity or sallies and points of wit in their poesies, yet there is some cadence and number in their verses, and some kind of rhyme in them. For an instance of which I join hereto a Greenland song, or ode, composed by one of the natives, who formerly lived in our colony, by name Frederick Christian, upon the birth day of his then royal highness, Prince Christian, on the 30th of November, 1729, which is as follows.

A GREENLAND SONG,
COMPOSED BY FREDERICK CHRISTIAN,

A Native

Entry.	Amna aja aja, aja aja, &c
One morning as I went out, and saw,	Annigamma irsigeik, amna aja aja, &c.
	Arvallirsullitlarmeta: amna aja, &c.
That flags and colours were flying,	Opellungarsullarmeta, amna aja, &c.
And that they made ready	Erkaiseigamig og, amna aja aja, &c.
To fire the guns;	Tava orkarbigeik, amna aja aja, &c.
Then I demanded,	Saag erkaisovise? Amna aja, &c.
Why do you fire?	Tava akkyanga, assuog Nellermago,
And they answered me, because	Okuine annivine nellermago, amna aja, &c.
the King's Son's	Angune tokkopet kongingoromagame, amna aja, &c.
Birth day was celebrated,	
Who is to be king after his father,	
And succeed in the kingdom.	Kingoreis semmane; amna aja, &c.
Thereupon I said to my friend,	Tava ikkinguntiga; amna aja aja, &c.
Let us make a song	Pitsimik sennegiluk; amna aja, &c.

To the King's Son;

For he shall be made king.

This my little song shall praise him:

'Tis said, he is a brave prince,

Let us therefore rejoice;

For he shall be our king,

After his Father's death,

We rejoice also, because

He loves us as his Father does;

Who sent over clergymen to us,

To teach us the word of God;

Lest we should go to the Devil.

Be thou like him, so shall we love thee,

And cherish thee,

And be thy servants.

Our ancestors have also been thy servants,

Even they.

That thou hast thought on us,

This we know very well, O gracious Son of the King.

We hope thou wilt continue so to do,

The King thy father has before possessed us,

When thou shalt be our King thou'lt prove good enough.

Whatever we possess

Shall be thine altogether.

When Greenland shall have received instruction,

Kongib imna niamganut, amna aja, &c.

Kongingoromamet; amna aja aja, &c.

Pisingvoara una; amna aja aja, &c.

Ostantigirfaræt sillakartok unnertlugo, amna aja, &c.

Tipeitsutigeik: amna aja aja, aja aja.

Kongingoromamet; amna aja aja, aja, &c.

Angune-oy tokkoppet: amna aja aja, &c.

Tipeitsokigogut: amna aja aja, aja, &c.

Attatatut asseigalloäpatit: amna aja, &c.

Pellesille tamaunga innekaukit: amna aja, &c.

Gudimik ajokarsokullugit: amna aja, &c.

Torngarsungmut makko inneille pekonnagit: amna aja, &c. Iblile tameitit neglitsomapaukit,

Asseigomarpaukit: amna aja aja, &c.

Kivgakomarpautigut: amna aja aja, &c.

Siurlit karalit kivgarimiaukit,

Juko: amna aja aja, aja aja, &c.

Isumatigautigut: amna aja aja, &c.

Nellungikallorapagut, Kongib Niarnga ajungitsotit, Teimatoy isumariotit: amna aja aja, &c.

Kongib Angutit pekaramisigut,

Iblile Kongingoruit namaksimotit: amna aja, &c.

Tomasa pirsaugut: amna aja aja, &c.

Piarmapotit makko: amna aja, &c.

Karalit illerpeta: amna aja, &c.

Gud negligomaparput, Kongible nalleklugo: amna aja aja, &c.

Tecpeitsukigisa: amna aja aja, aja, &c.

Kongiblo Niarnga: amna aja aja, &c.

Then shall they love God and
honour the King.
Let us be merry,
And of the King's Son
Drink the health.
And say, Long live Christian!
And thy Consort.
May thy years be many!
(This I wish) Frederick
Christian, and my friend
Peter, who were the first
baptized of Greenland.
Would to God our countrymen
were also.

Skaalia immerlugo: amna aja aja, &c.
Tave okarpogut, Christian innuvit: amna
aja, &c.

Nulliello: amna aja aja, aja aja, &c.
Okiutikit armarlesorsuangorlutik: amna
aja, &c.
Friderik Christian ikingutigalo; amna, &c.
Peder, karalinit kockkartoguk: amna, &c.
Kannoktok! Ekkarlivut tamakilit makko:
amna aja, &c.
Amna, aja aja, aja aja, aja aja, hei!

They have, besides this, another sort of diversion, accompanied with singing, which consists in swopping or bartering. He that performs the office of drummer and singer, exposes one thing or other to sale, at any rate he thinks fit; if any of the company has a liking to it, he shows his consent by giving the seller a slap on his breech, and the bargain is done, and cannot be retrieved, whether good or bad. The boys and lads have also their pastimes and plays, when they meet in the evening. They take a small piece of wood, with a hole in it at one end, to that they tie a little pointed stick with a thread or string, and throwing the piece with the hole in it up into the air, they strive to catch it upon the pointed stick, through the hole. He that does it twenty times successively, and without failing, gains the match, or party, and he that misses gets a black stroke on his forehead for every time he misses. Another boy's play is a game of chance, like cards or dice; they have a piece of wood pointed at one end, with a pin or peg in the midst, upon which it turns; when the boys are seated around, and every one laid down what they play for, one of them turns the pointed piece of wood with his finger, that it wheels about like a mariner's compass; and when it has done, he that the point aims at, wins all that was laid down. Ball playing is their most common diversion, which they play two different ways. They divide themselves into two parties; the first party throws the ball to each other; while those of the second party endeavour to get it from them, and so by turns. The second manner is like our playing at football. They mark out two barriers, at three or four hundred paces distance one from the other; then being divided into two parties, as before, they meet at the starting place, which is at the midway between the two barriers; and the ball being thrown upon the ground, they strive who first shall, get at it, and kick it with the foot, each party towards their barrier. He that is the most nimble footed and dextrous at it, kicking the ball before him, and getting the first to the barrier, has won the match.

Thus (they will tell you) the deceased play at football in Heaven, with the head of a morse, when it lightens, or the North-light (aurora borealis) appears, which they fancy to be the souls of the deceased.

When their acquaintance from abroad come to see them, they spend whole days and nights in singing and dancing; and as they love to pass for men of courage and valour, they will try forces together, in wrestling, struggling, and playing hook and crook, which is to grapple with the arms and fingers made crooked, and intangled like hooks. Whoever can pull the other from his place, thinks himself a man of worth and valour. The women's or rather the maiden's plays, consist in dancing around, holding one another by the hand, forming a circle, and singing of songs.

Chapter 16

OF THEIR LANGUAGE

Though the Greenland language has not affinity with other European tongues, yet it seems to have borrowed some words from the Norwegians, who formerly inhabited part of the land; for such words agree both in name and signification; as, for example, *Kona*, a Woman; *Nerriok*, to eat, from the Norway word *Noerrie*. The herb *Angelica*, which they in Norway call *Quaun*, the Greenlanders call *Qvaunnek*. A Porpoise, in Norway called *Nise*, they call *Nise*. Ashes, in Norway, *Aske*, in Greenland, *Arkset*. A Lamp, in the Norwegian, *Kolle*, in the Greenlandian, *Kollek*. Some of their words resemble Latin words of the same signification; as, *Gutta*, a drop; in the Greenland tongue, *Gutte*, or *Kutte*. *Ignis*, Fire, they call *Ingnek*. And some they have got from Hebrew roots, as, *Appa*, a word the children use to name their father, and some others.

The accent and pronunciation of it is hard and difficult, because they speak very thick, and in the throat. The same language is spoke throughout the whole country, though the accent and pronunciation differs here and there as different dialects; chiefly towards the Southern parts, where they have received and adopted many foreign words, not used in the Northern parts. But the angekuts, or divines, make use of a particular speech, whenever they conjure; for then they use metaphorical locutions and words in a contrary sense. The women-kind also have a particular pronunciation peculiar to themselves, and different from that of the men, making use of the softest letters at the end of words, instead of hard ones; for example, *Am* for *Ap*, that is, *Yes. Saving*, for *Savik*, a *Knife*. Their language, in common, wants the letters, *c, d, f, q, x*. They have besides many double and unknown consonants, which is the cause, that many of their words cannot be spelt according to their manner of pronouncing them. For the rest, their expressions are very natural and easy, and their constructions so neat and regular, that one would hardly expect so much from a nation so unpolite and illiterate. The language is very rich of words and sense, and of such energy, that one is often at a loss and puzzled to render it in Danish; but then again it wants words to express such things as are foreign, and not in use among them. They have monosyllables and polysyllables, but most of the last. Their words, as well nouns as verbs, are inflected at the end, by varying the terminations, without the help of the articles or particles, like the Greek and Latin. The adjectives always follow their substantives; but the possessive pronouns are joined to the nouns, as the Hebrew suffix a:[35] nor have the nouns alone their suffixa, but the verbs also. To satisfy the reader's curiosity, I have hereto joined a list of some of the words and a sketch, showing the construction and inflections of this language.

VOCABULARY OF THE LANGUAGE OF GREENLAND

Singular.	Dual.	Plural.
Innuk, Mankind,	Innuk,	Innuit.
Angut, a Man,	Angutik,	Angutit.
Arnak, a Woman,	Arnek,	Arnet.
Niakok, the Head,	Niakuk,	Niakut.
Irse, an Eye,	Irsik,	Irsit.
Kingak, the Nose,	Kingek,	Kinget.

---

[35] In its inflections it agrees with the Hebrew.

Kinak, the Face,	Kinek,	Kinet.
Kannek, Mouth,	Kannek,	Kangit.
Okak, Tongue,	Okek,	Oket.
Kiut, a Tooth,	Kiutik,	Kiutit.
Kartlo, a Lip,	Kartluk,	Kartluit.
Suit, an Ear,	Siutik,	Siutit.
Nyak, Head of Hair,	Nytkiek,	Nytkiet.
Sækik, the Breast,	Sækkirsek,	Sækkirset.
Iviange, Bubby,	Iviangik,	Iviangit.
Tue, Shoulder,	Tubik,	Tubit.
Tellek, Arm,	Tellik,	Tellit.
Ikusik, Elbow,	Ikivtik,	Ikivtit.

Arkseit, Hand (that is the Fingers), is plural only.

Tikek, Finger,	Tikik,	Tirkerit.
Kukik, Nail,	Kukik,	Kuket.
Nak, Belly,	Nersek,	Nerset.
Innelo, Bowel,	Inneluk,	Inneluit.
Okpet, the Thigh,	Okpetik,	Okpetit.
Sibbiak, the Hip,	Sibbirsek,	Sibbirset.
Serkok, Knee,	Serkuk,	Serkuit.
Kannak, Shank,	Kannek,	Kannerset.

Isiket, Foot, is only of the plural number.

Kimik, Heel,	Kimik,	Kimikt.

The construction with Possessive Pronouns is thus.

Iglo, *a House*,	Igluk,	Iglut.	
My House,	Igluga,	Igluka,	Igluka.
Thy House,	Iglut,	Iglukit,	Iglutit.

His House,	Igloa,	Igluk,	Igloëi.
His own House,	Iglune,	Iglugne,	Iglune.
Our House,	Iglout,	Iglogat,	Iglovut.
Your House,	Iglurse,	Iglursik,	Igluse.
Their House,	Igloæt,	Igloæk,	Iglöeit.
Their own House,	Iglurtik,	Iglutik,	Iglutik.

This same Noun's construction with the suffixas at Prepositions, *mik* and *nik*, *mit* and *nit*, which signifies from; *mut* and *nut*, to; *me* and *ne*, on or upon, is thus performed.

Singular.	Dual.	Plural.	
To the House,	Iglomut,	Iglugnut,	Iglunut.
To my House,	Iglumnut,	idem,	idem.
To thy House,	Iglungnut,	idem,	idem.
To his House,	Igloanut,	Igloennut,	Iglocinut.
To his own House,	Iglominut,	Iglungminut,	Iglominut.
To our House,	Iglotivnut,	Iglutivnut,	idem.
To your House,	Iglusivnut,	idem,	idem.
To their House,	Igloænut,	idem,	Iglöeinut.
To their own House ,	Iglomingnut,	idem,	idem.

As to the verbs, they are either simple or compounded: there are five conjugations, to which may be added a sixth of negative verbs. There are three tenses in all, the present, preterit, and future; and six moods, *viz.* indicative, interrogative, imperative, permissive, conjunctive, and infinitive.

The examples of the simple verbs are these. The first conjugation ends in *kpok*, as *Ermikpok*, he washes himself: *Aglekpok*, he writes.

The second ends in *rpok*, as *Mattarpok*, he undresses himself: *Aularpok*, he sets out on a journey: *Ajokarsorpok*, he teaches.

The third conjugation ends in *pokpurum*; that is, in *pok* preceded by a vowel, as *Egipok*, he throws away; *Inginok*, he sits down; *Akpapok*, he runs.

The fourth ends in *ok* or *vok*, as *Pyok*, he receives: *Aglyok*, he grows: *Assavok*, he loves.

The fifth conjugation ends in *au*, as *Irsigau*, he ogles; *Arsigau*, he resembles; *Angekau*, he is tall.

The sixth conjugation of negative verbs ends in *ngilak*, as *Ermingilak*, he does not wash himself: *Mattengilak*, he does not undress himself: *Pingilak*, he receives not: *Egingilak*, he throws not away: *Irsigingilak*, he ogles not.

Inflexion of a Verb with the suffixes of a person agent of the first conjugation in *kpok*.

Indicative.		Present.
Singular.	Dual.	Plural.
He washes himself,	*The two wash themselves,*	*They wash themselves,*
Ermikpok.	Ermikpuk.	Ermikput.
I wash myself,	*We two wash ourselves,*	*We wash us.*
Ermikpunga.	Ermikpoguk.	Ermikpogut.
Thou wash thyself,	*You two wash yourselves,*	*You wash yourselves.*
Ermikpotit.	Ermikpotik.	Ermikpose.

The inflexion with suffixes of a person patient is formed this way.

Thou washest me.	*Ye two wash me,*	*You wash me,*
Ermikparma,	Ermikpautiga.	Ermikpausinga.
He washes me,	*The two wash me,*	*They wash me,*
Ermikpanga.	Ermikpainga.	Ermikpanga.
I wash him,	*We two wash him,*	*We wash him,*
Ermikpara.	Ermikparpuk.	Ermikparput.

He washes him,	*The two wash him,*	*They wash him,*
Ermikpæ.	Ermikpæk.	Ermikpæt.
Thou washest him,	*Ye two wash him,*	*You wash him,*
Ermikpet.	Ermikpartik.	Ermikparse.
I wash thee,	*We two wash thee,*	*We wash thee,*
Ermikpaukit.	Ermikpautikit.	Ermikpæutigit.
He washes thee,	*The two wash thee,*	*They wash thee,*
Ermikpatit.	idem.	idem.
Thou washest us,	*Ye two wash us,*	*You wash us,*
Ermikpautigut.	——pautigut.	Ermikpausigut.
He washes us,	*The two wash us,*	*They wash us,*
Ermikpatigut.	idem.	idem.
I wash you,	*We two wash you,*	*We wash you,*
Ermikpause,	idem.	idem.
He washes you,	*The two wash you,*	*They wash you,*
Ermikpase.	idem.	idem.
I wash them,	*We two wash them,*	*We wash them,*
Ermikpaka.	Ermikpauvut.	idem.
He washes them,	*The two wash them,*	*They wash them,*
Ermikpei.	Ermikpatik.	Ermikpase.
Thou washest them,	*Ye two wash them,*	*Ye wash them,*
Ermikpatit.	Ermikpatik.	Ermikpeit.

Inflexion of the Negative Verb.

He washes not	*The two wash not*	*They wash not*
himself,	*themselves.,*	*themselves.,*
Ermingilak.	Ermingilek.	Ermingilat.

I do not wash myself,	*We two wash not ourselves,*	*We wash not ourselves,*
Ermingilanga.	Ermingilaguk.	Ermingilagut.
Thou dost not wash thyself,	*Ye two do not wash yourselves,*	*You do not wash yourselves,*
Ermingilatit.	Ermingilatik.	Ermingilase.

With the suffixes of the patient person the negative verbs are inflected like the affirmatives; as,

He washes me not,	*Ye two wash me not,*	*They wash me not,*
Ermingilanga.	idem.	idem.
Thou washest me not,	*Ye two wash me not,*	*You wash me not,*
Ermingilarma.	Ermingilautinga.	Ermingilausinga.

And in the same manner you may inflect all verbs whatsoever.

The preterits and futures have the same suffixa as the present tense.

Concerning the compounded verbs, it is to be observed, that, whereas their auxiliary verbs are but few, they make use of several particles to supply their place, which are annexed to the simple verbs, and so make them compounded verbs, yet these particles by themselves are not used, nor of any signification. And by this connection or composition the simple verbs change their conjugation. As for example,

First, in this expression, they used to do so and so, the composition is formed thus; of the simple verb *Erminpok*, he washes himself, in the composition is made *Ermingarace*, he uses to wash himself. *Kieavok*, he weeps; *Kieeillarau*, he uses to weep; *Aularpok*, he goes from home; *Aulararau*, he uses to go from home.

Second, when the expression runs thus, he comes to do this or that, it is turned in this manner. *Ermigiartorpok*, he comes to wash himself; *Aglegiartorpok*, he comes to write. And so in all other compositions.

But there are not only verbs compounded with one, but sometimes with two, three, or more particles joined to the verb, when there is a longer sentence to be expressed. And for this reason, the words and particles undergo a great many changes and variations, inasmuch as they retain but certain radical letters, the rest either being thrown away and quite lost, or else changed for others. As for instance, *Aulisariartorasuarpok*, he made haste to go out a fishing. Here three verbs are joined together in one. *Aulisarpok*, he fishes; *Peartorpok*, to go about something; and *Pinnesuarpok*, to make haste. Again, *Aglekkinniarit*, endeavour to write better. Here we have another threefold composition. First, *Aglekpok*, he writes; then *Pekipok*, to mend, or do better, and at last *Pinniarpok*, to endeavour. From whence comes the verb *Aglikkinniarpok*, he endeavours to write better; in the imperative mood, *Aglekkinniarit*, as above.

The Creed, and the Lord's Prayer, Translated into the Greenland Language

Article I

Operpunga   Gud-mun   Attatavnut,   ajuakangitsomut,   killagmik nunamiglo sennarsomut.

Article II

Operpunga Jesus Christusmut, Ernetuanut, Nallegautimut, Annersamit helligmit pirsok, Niviarsamit Mariamit erniursok; anniartok Pontius Pilatus-mit; Isektitaursok, tokkorsok, illirsorto, allernum akkartok. Ullut pingajuane tokkorsonit makitok; Killangmut Kollartok; Angume Gub tellerpiet tungane ipsiarsok; tersanga amma tikiytsomaryok, umarsullo tokongarsullo auiksartitsartorlugit.

Article III

Operpunga Gub Annersanut, opertokartoniglo nuname: Innungliglo helligniglo illegeinik, Synderronermiglo, Timiniglo umaromartonik, tokkorsublo Kingorna tokkoviungitsokartomik. Amen.

The Lord's Prayer.

NALLEKAM OKAUSIA.

Attavut killangmepotit, akkit usorolirsuk; Nallegavet aggerle; pekorset Killangme nunam etog tamaikile: Tunnisigun ullume nekiksautivnik; pissarauneta aketsorauta, pisingilaguttog akectsortivut; Ursennartomut pisitsaraunata; ajortomin annautigut: Nallegauet, Pisarlo, usornartorlo pigangaukit isukangithomun. Amen.

Chapter 17

OF THE GREENLAND TRADE, AND WHETHER, IN PROMOTING IT, THERE IS ANY ADVANTAGE TO BE EXPECTED

The goods and commodities Greenland affords for the entertaining of commerce, or traffic, are whale blubber or fat, and whale bones, unicorn horns, rein deer skins and hides, seal and fox skins. These wares they barter against merchandizes of our produce, as coats and shirts made of white, blue, red or striped linen or woollen cloth; as also knives, hand-saws, needles, hooks to angle with, looking-glasses, and other such merchandize or hardwares: besides what they buy of wood, as rafts, poles, deal boards, chests; and of brass and copper, as kettles and the like, tin dishes and plates; for which they pay to the full price. At the beginning of our late settlement in those parts the trade was much brisker than at present, and much more profitable; for foreign traders flocking thither in great numbers have so overstocked them with goods, and undersold one another, to draw the natives to them from others, that the trade is considerably slackened and fallen. Yet I trust, that, if we once became masters of this trade, as it in justice belongs to us, by the right the King of Denmark lawfully claims to these countries as much as any kingdom or province subject to him; I trust, that, with this proviso, the trade to

Greenland would prove as profitable as any other whatsoever; which has been evidenced not long ago, when by his Majesty's special order foreign trade has been prohibited within a certain distance on each side of the colonies. For if the lading of some ships with fish and train from Finmark, and others of fish, train, salt meat, and butter from Iceland and Fero, bring to the traders considerable profit; who would question, but the same or better advantage may be expected from the importing quantities of whale train, whale bones, rein deer hides, fox and seal skins, which are of more value than the Iceland or Feroe? And, if the produce or commodities of Greenland were formerly reckoned of that importance, that they were deemed sufficient to maintain the King's table, why not also at present? provided Greenland may by settlements and improvement retrieve its former abundance, which is not impossible.

If the old lands, formerly inhabited and manured by the Norway colonies, were anew peopled with men and cattle; they would, without doubt, yield as much as either Iceland or Feroe, seeing there is as good pasture ground as in those islands. I shall forbear to mention salmon and cod fishing, as it seems at present to be but of little or no importance, especially on the West side; though I am credibly informed by the natives, that on the Southern coast they catch abundance of fine large cod. Yet this may be more than sufficiently compensated by the whale fishery on the North and the capture of seals on the South, which if rightly undertaken, and with vigour set on foot, will bring as much, nay far more profit than the salmon and cod catching does in other places; chiefly the seal capture, which can be undertaken at very small expenses, *viz.* at the coast with strong nets, with which they may catch many thousands in Greenland; which, if hitherto not practised, ought to be imputed to negligence and want of a good regulation. In short, Greenland, as we see, is very convenient for trading, and may be very well worth one's while to take in hand. But there is little to be done, without an established and formed company of men of substance as well as resolution; being altogether impossible and above the strength of any private man to master it and go through with it.

Chapter 18

THE RELIGION, OR RATHER SUPERSTITION, OF THE GREENLANDERS

The Greenlanders' ignorance of a Creator would make one believe they were atheists, or rather naturalists. For, when they have been asked from whence they thought that Heaven and Earth had their origin, they have answered nothing, but that it had always been so. But if we consider, that they have some notion of the immortality of souls,[36] and that there is another much happier life after this; moreover, as they are addicted to different kinds of superstition, and that they hold there is a Spiritual Being, which they call Torngarsuk, to whom they ascribe a supernatural power, though not the creation or the production of creatures (of whose origin they tell many absurd and ridiculous stories), all this, I say, supposes some sort of worship; although they do not themselves, out of their brutish stupidity, understand or infer so much, or make use of the light of nature and the remaining spark of the image of God in their souls, to consider the invisible being of God by his visible works, which is the creation of the world.—Rom. i. For which reason, instead of attaining the knowledge of

---

[36] The angekuts say that souls are a soft matter to feel, or rather that they cannot be felt, as if they had neither sinews nor bones.

God and true religion, they are unhappily fallen into many gross superstitions.

But notwithstanding that all these superstitions are authorized by, and grounded upon the notion they have of him they call Torngarsuk, whom their lying angekuts or prophets hold for their oracle, whom they consult on all occasions, yet the commonalty know little or nothing of him, except the name only: nay even the angekuts themselves are divided in the whimsical ideas they have formed of his being; some saying he is without any form or shape; others giving him that of a bear, others again pretending he has a large body and but one arm; and some make him as little as a finger. There are those who hold he is immortal, and others, that a puff of wind can kill him. They assign him his abode in the lower regions of the Earth, where they tell you there is constantly fine sunshiny weather, good water, deer, and fowls in abundance. They also say he lives in the water; wherefore, when they come to any water, of which they have not drank before, and there be any old man in the company, they make him drink first, in order to take away its Torngarsuk, or the malignant quality of the water, which might make them sick and kill them. They hold furthermore, that a spirit resides in the air, which they name Innertirrirsok, that is, the Moderator or Restrainer, because it is pursuant to his order, that the angekuts command the people to restrain or abstain from certain things or actions, that they may not come into harm's way. According to their theology, or mythology, there is yet one spirit, harbinger of the air, whom they stile Erloersortok, which signifies a Gutter, because he guts the deceased, and feeds upon their intestines. His countenance, they say, is very ghastly and haggard, hollow eyes and cheeks, like a body that is starved.

Each element has its governor or president, which they call Innuæ;[37] from whence the angekuts receive their torngak, or familiar spirits. For

---

[37] The Innuæ, or inhabitants of the sea, they call Kongeuserokit; of whom they say, that they feed upon fox tails. Ingnersoit, a sort of sea sprites, which inhabit the rocks that lie upon the coast; which, they tell you, will carry away the Greenlanders, not to do them any harm, but to enjoy their company. Tunnersoit are phantoms living in the mountains; and Ignersoit, or fiery sprites (because they appear to be all over fire) live near the shore, in steep and craggy cliffs. This is that meteor which we call the Flying Dragon. Innuarolit they pretend to be a

every angekkok has a torngak, who attends him, after he has ten times conjured in the dark.

Some have their own deceased parents for their torngak, and others get theirs out of some of our nation, who they say discharge their fire arms when they wait before the entry of the place where the angekkok performs his conjuration. Whether Torngak and Torngarsuk be one and the same thing I shall not decide; but certain it is, that one is derived from the other. From Torngarsuk the angekuts pretend they learn the art of conjuring; which they are taught in this method. If one aspires to the office of an angekkok, and has a mind to be initiated into these mysteries, he must retire from the rest of mankind, into some remote place, from all commerce; there he must look for a large stone, near which he must sit down and invoke Torngarsuk, who, without delay, presents himself before him. This presence so terrifies the new candidate of angekutism, that he immediately sicken, swoons away, and dies; and in this condition he lies for three whole days; and then he comes to life again, arises in a newness of life, and betakes himself to his home again. The science of an angekkok consists of three things. 1. That he mutters certain spells over sick people, in order to make them recover their former health. 2. He communes with Torngarsuk, and from him receives instruction, to give people advice what course they are to take in affairs, that they may have success, and prosper therein. 3. He is by the same informed of the time and cause of any body's death; or for what reason anybody comes to an untimely and uncommon end; and if any fatality shall befal a man. And though this lying spirit of the angekuts is oftentimes found out by their gross mistakes, when the events do not answer their false predictions, as commonly happens; yet, for all that, they are in great honour and esteem among this stupid and ignorant nation, insomuch that nobody ever dare refuse the strictest obedience to

---

people of a dwarfish size, like pigmies, and are said to inhabit the East side of Greenland. Erkiglit, on the contrary, are said to be a nation of a huge and monstrous size, with snouts like dogs; they are likewise said to dwell on the East side. Sillagiksortok, a spirit, who makes fair weather, and lives upon the ice mountains. Nerrim Innua, or the ruler of diet, because he prescribes rules for the diet or eating of those that are obliged to keep abstinence. They ascribe also some sort of divinity to the air, and for fear of offending it they will refrain from certain things and actions; for which reason they are afraid to go out in the open air in the dark.

what they command him in the name of Torngarsuk, fearing, that, in case of disobedience, some great affliction and misfortune may happen to him. Among many other fibs, and most impudent lies, they make also these silly stupid wretches believe, that they can, with hands and feet tied, mount up to Heaven, and see how matters stand there; and likewise descend to Hell, or the lower regions of the Earth, where the fierce Torngarsuk keeps his court. A young angekkok must not undertake this journey but in the fall of the year, by reason, that then the lowermost Heaven, which they take the rainbow to be, is nearest to the Earth.

The farce or imposture is thus acted: a number of spectators assemble in the evening at one of their houses, where, after it is grown dark, everyone being seated, the angekkok causes himself to be tied, his head between his legs and his hands behind his back, and a drum is laid at his side; thereupon, after the windows are shut and the light put out, the assembly sings a ditty, which, they say, is the composition of their ancestors; when they have done singing the angekkok begins with conjuring, muttering, and brawling; invokes Torngarsuk, who instantly presents himself, and converses with him (here the masterly juggler knows how to play his trick, in changing the tone of his voice, and counterfeiting one different from his own, which makes the too credulous hearers believe, that this counterfeited voice is that of Torngarsuk, who converses with the angekkok.) In the mean while he works himself loose, and, as they believe, mounts up into Heaven through the roof of the house, and passes through the air till he arrives into the highest of heavens, where the souls of angekkut poglit, that is, the chief angekkuts, reside, by whom he gets information of all he wants to know. And all this is done in the twinkling of an eye.

Concerning the angekkut poglit, whom we just now mentioned, as they pass for the heads of the clergy, and are reckoned the most eminent and wisest of all, they also must pass through the inferior orders, and several hard trials, before they can attain to this high degree of pre-eminency; for none is deemed worthy of such a dignity, but he that has made his noviciateship in the lower rank, as an ordinary angekkok. The trial he must undergo, is this: they tie his hands and feet, as aforesaid, and after the light is put out, and they are all left in darkness (that nobody may see how the trick is played, and their imposture be discovered), then they pretend that a white bear enters the room, takes hold of his great toe with his teeth, and dragging him along to the sea shore, jumps with him into the sea, where a morse is ready, and takes hold of him by his privy parts, devouring him, together with the white bear. A little while after all his bones are thrown in upon the floor, one after another, not one missing; and then his soul rises up off the ground, which gathers the bones, and animates the whole body again, and up starts the man, a hale and entire as ever he was; and thus he is made an angekkok poglik.

The angekkuts, as before observed, are kept in great honour and esteem, and beloved and cherished as a wise and useful set of men; they are also well rewarded for their service, when it is wanted. But, on the contrary, there is another sort of conjurers or sorcerers, especially some decrepid old women, which they call illiseersut, or witches, who persuade themselves and others, that, by the virtue of their spells and witchcraft they can hurt people in their life and goods. These are not upon the same footing with the angekkuts; for as soon as any one incurs only the suspicion of such demeanor, he or she is hated and detested by everybody, and at last made away with, without mercy, as a plague to mankind, and not deemed worthy to live.

Moreover the angekkuts abuse the people's credulity, making them believe, that they can cure all sorts of diseases; though they apply such remedies as have no virtue in them to cure, such as muttering of spells, and blowing upon the sick bodies; wherein they resemble to a hair those conjurers of which the prophet Isaiah speaks, chapter viii, verse 19.

And if by chance any one, who has been under these jugglers' hands, recovers, they do not fail to ascribe it to the virtue of their juggling tricks. At times they use this way of curing the sick; they lay him upon his back, and tie a ribbon, or a string, round his head, having a stick fastened to the other end of the string, with which they lift up the sick person's head from the ground, and let it down again; and at every lift he communes with his Torgak, or familiar spirit, about the state of the patient, whether he shall recover or not; now, if his head is heavy in lifting it, it is with them a sign of death; if light, of recovery.[38] Notwithstanding all this, I am loth to believe, that, in these spells and conjurings, there is any real commerce with the devil; for to me it clearly appears, that there is nothing in it but mere fibs, juggling tricks, and impostures, made use of by these crafty fellows for the sake of filthy lucre, for they are well paid for their pains. Nevertheless, it cannot be denied, but that the evil spirit has a hand in all this, and is the chief actor upon this stage, to keep these poor wretches in their chains, and hinder them from coming to the true knowledge of God.

The angekkuts can also persuade whom they please, that they have no souls, especially if they are in a bad state of health, pretending they have the power to create new souls in them, provided they pay them well for it, which the ignorant fools are very willing to do. They prescribe to all rules of conduct and behaviour in different cases, which rules none dare refuse to live up to with the greatest exactness imaginable; as for example, if any dies in a house, those of the house cannot, for a set time, do all sorts of work; especially the relations of the deceased are obliged to abstain, not only from certain works, but likewise from certain victuals.

If a patient be under the hands of an angekkok, he must live by rule, which they are accustomed to observe so exactly, that even when we have assisted many of them with our medicaments, they have always demanded what sort of diet they were to keep. Women in childbed are to abstain from working, and from certain victuals, viz. flesh meat, which their own husbands have not taken, or that of a deer, whose entrails are not sound, but damaged. The first week after the delivery they eat nothing but fish, afterwards they are allowed meat. The bones they pick in this state must not be carried out of doors. After the first childbed, a woman is not allowed to eat of the head or liver. They must not eat in the open air. During their lying-in they have their water pails for themselves alone; if any unwittingly should drink of this water, the rest must be thrown away. Their husbands must forbear working for some weeks, neither must they drive any trade during that time: likewise if anybody be sick, they do not care to meddle

---

[38] While angekkuts are conjuring, nobody must scratch his head, nor sleep, nor break wind; for they say, that such a dart can kill the enchanters, nay the devil himself. After a conjuration has been performed, there is a vacancy from working for three or four days.

with any trade. They are not allowed to eat or drink bareheaded. They pull off one of their boots, and lay it under the bowl which they eat out of, to the end (as they imagine) that the infant, being a male, may become a good seal catcher. During the infancy of the child, they dare not boil anything over the lamp, nor let any strangers light a fire with them; and many more fooleries to be observed.[39] It is customary among them for married women to wash and cleanse themselves after their months, that their husbands may not catch a distemper and die. Likewise, if they have happened to touch a dead corpse, they immediately cast away the clothes they have then on; and for this reason they always put on their old clothes when they go to a burying, in which they agree with the Jews, as in many other usages and ceremonies; for example, to bewail the loss of their virginity; to mark themselves upon their skin; to cut their hairs round the head, which the Lord forbids the Jews to do, Levit. xix. When I consider this and many other of their customs, which seem to be of a Jewish extraction, I am not far from acceding to the opinion of a certain famous writer, concerning the Americans; among whom as he found sundry Jewish rites and ceremonies, he took them to descend from Jews, or rather from some of the ten tribes of Israel, who were led into the Assyrian captivity, and afterwards dispersed into unknown countries.—See hereon Espars, 1. iv.

A superstition very common among them is, to load themselves with amulets or *pomanders* dangling about their necks and arms, which consist of some pieces of old wood, stones or bones, bills and claws of birds, or anything else, which their fancy suggests to them; which amulets, according to their silly opinion, have a wonderful virtue to preserve those that wear them from diseases and other misfortunes, and gives them luck to good captures. To render barren women fertile or teeming, they take old pieces of the soles of our shoes to hang about them; for, as they take our nation to be more fertile, and of a stronger disposition of body than theirs, they fancy the virtue of our body communicates itself to our clothing.

---

[39] *Argnakaglertoko*, a woman that lives by rule, they say, can lay the storm, by going out of doors and filling her mouth with air, and coming back into the house, blows it out again. If she catches the rain drops with her mouth, it will be dry weather; and other strange effects they ascribe to her.

Concerning the creation and origin of all things, they have little to say, but they think all has been as it ever will be. Nevertheless they abound in fables in regard to these matters. Their tale of the origin of mankind runs thus: at the beginning one man, *viz.* a Greenlander, sprung out of the ground, who got a wife out of a little hillock.[40] From these are descended lineally the Greenlanders; which may pass for a remnant, though an adulteration from the true tradition of the origin of man. But as to us foreigners, whom they stile *Kablunæt* (that is, of a strange extraction), they tell a most ridiculous story, importing our pedigree from a race of dogs; they say, that a Greenland woman once being in labour, brought forth at the same time both children and whelps: these last she put into an old shoe, and committed them to the mercy of the waves, with these words; Get ye gone from hence and grow up to be Kablunæts. This, they say, is the reason, why the Kablunæts always live upon the sea; and the ships, they say, have the very same shape as their shoes, being round before and behind.

The reason why men die, they tell us, is, that a woman of their nation once uttered these words; *Tokkolarlutik okko pillit, sillarsoak rettulisavet*, Let them die one after another; for else the world cannot hold them. Others relate it in this manner: two of the first men contended with one another, one said, *Kaut sarlune unnuinnarluna, innuit tokkosarlutik*; that is, Let there be day, and let there be night, and let not men die. The second said, *Unnuinnarlune, kausunane, innuit tokkosinnatik*; that is, Let there be nothing but night, and no day, and let men live; and after a long contention the first saying got the day. Of the origin of fishes and other sea animals they tell a ridiculous story, *viz.* an old man was once cutting chips off of a piece of wood; with these chips he rubbed himself between the thighs, and threw them into the sea, whereupon they immediately became fishes. But of a certain fish called hay, they derive his production from this accident, that a woman washing her hairs in her own water, a blast of wind came and carried away the clout with which she dried her hairs, and out of that clout

---

[40] A word not known to me in the Danish tongue.

was produced a hay fish; and for this reason they say, the flesh of this fish has got the smell of urine.

They have got no notion of any different state of souls after death; but they fancy that all the deceased go into the land of the souls, as they term it. Nevertheless they assign two retreats for departed souls, *viz.* some go to Heaven, others to the centre of the Earth; but this lower retirement is in their opinion the pleasantest, inasmuch as they enjoy themselves in a delicious country, where the sun shines continually, with an inexhaustible stock of all sorts of choice provision. But this is only the receptacle of such women as die in labour, and of those that, going a whale fishing, perish at sea; this being their reward, to compensate the hardships they have undergone in this life; all the rest flock to Heaven.

In the centre of the Earth, which they reckon the best place of all, they have fixed the residence of Torngarsuk and his grandame, or (as others will have it) his lady daughter, a true termagant and ghastly woman, to whose description, though already made in my continuation of the relations of Greenland, some time ago published, I shall yet allow a place in this treatise, and is as follows. She is said to dwell in the lower parts of the earth under the seas, and has the empire over all fishes and sea-animals, as unicorns, morses, seals, and the like. The bason placed under her lamp, into which the train oil of the lamp drips down, swarms with all kinds of sea fowls, swimming in and hovering about it. At the entry of her abode is a *corps de garde* of sea dogs, who mount the guard, and stand sentinels at her gates to keep out the crown of petitioners.[41] None can get admittance there but angekuts, provided they are accompanied by their Torngak, or familiar spirits, and not otherwise. In their journey thither they first pass through the mansions of all the souls of the deceased, which look as well,

---

[41] Others say, that a huge dog watches the entry, and gives warning, when an angekkok attempts to get in, and defends the entry. Wherefore the angekkok must watch the minute, that the dog falls asleep (which lasts but a moment), to steal in upon her. This moment nobody knows but an angekkok poglik; wherefore the other angekkuts often return home again without success. This frightful woman is said to have a hand as big as the tail of a whale, with which, if she hits anybody, he is at one stroke mouse-dead. But if the angekkok conquers her (which he does if he can get at her *aglerrutut*, which hang dangling about her face, and rob her of them) then she must discharge all fishes and sea animals, which she has detained in captivity; who thereupon return to their wonted stations in the sea.

if not better, than ever they did in this world, and want for nothing. After they have passed through this region, they come to a very long, broad, and deep whirlpool, which they are to cross over, there being nothing to pass upon but a great wheel like ice, which turns about with a surprising rapidity, and by the means of this wheel the spirit helps his angekkok to get over. This difficulty being surmounted, the next thing they encounter is a large kettle, in which live seals are put to be boiled; and at last they arrive, with much ado, at the residence of the devil's grandame, where the familiar spirit takes the angekkok by the hand through the strong guard of sea dogs. The entry is large enough, the road that leads is as narrow as a small rope, and on both sides nothing to lay hold on, or to support one; besides that, there is underneath a most frightful abyss or bottomless pit. Within this is the apartment of the infernal goddess, who offended at this unexpected visit, shows a most ghastly and wrathful countenance, pulling the hair off her head: she thereupon seizes a wet wing of a fowl, which she lights in the fire, and claps to their noses, which makes them very faint and sick, and they become her prisoners. But the enchanter or angekkok (being beforehand instructed by his Torngak, how to act his part in this dismal expedition) takes hold of her by the hair, and drubs and bangs her so long, till she loses her strength and yields; and in this combat his familiar spirit does not stand idle, but lays about her with might and main. Round the infernal goddess's face hangs the aglerrutit (the signification of which is to be found in my son's journals) which the angekkok endeavours to rob her of. For this is the charm, by which she draws all fishes and sea animals to her dominion, which no sooner is she deprived of, but instantly the sea animals in shoals forsake her, and resort with all speed to their wonted shelves, where the Greenlanders catch them in great plenty. When this great business is done, the angekkok with his Torngak proud of success make the best of their way home again, where they find the road smooth, and easy to what it was before.

As to the souls of the dead, in their travel to this happy country, they meet with a sharp pointed stone, upon which the angekkuts tell them they must slide or glide down upon their breech, as there is no other passage to get through, and this stone is besmeared with blood; perhaps, by this mystical or hieroglyphical image, they thereby signify the adversities and tribulations those have to struggle with, who desire to attain to happiness.

Chapter 19

THE GREENLANDERS' ASTRONOMY, OR THEIR THOUGHTS CONCERNING THE SUN, MOON, STARS, AND PLANETS

The notions the Greenlanders have of the origin of heavenly lights, as Sun, Moon, and Stars, are very nonsensical; in that they pretend that they have formerly been so many of their ancestors, who on different accounts were lifted up to Heaven and became such glorious celestial bodies.

Their silly stories concerning this matter have been related in the continuation to the Greenland Memoirs, or relations, but as this book very likely may not come to the hands of everybody, I shall shortly remember some of them here. The Moon, as they will have it, has been a young man, called Anningait, or Anningasina; whose sister was the Sun, named Malina, or Ajut (by which latter name they call any handsome woman, for whom they have a value, Ajuna.) The reason (why these two were taken up into Heaven) they give, is this: there were once a number of young men and women assembled to play together in a house made of snow (according to their custom in the winter season), when the Moon or Anningait, who was deeply in love with his sister, who assisted at this assembly, was used every night to put out the light, that he might caress her undiscovered; but she not liking these stolen caresses, once blackened her hands with soot, that she might mark the hands, face, and clothes of her unknown lover, who in the dark made addresses to her, and by that discover who he was: hence, they say, come the spots that are observed in the moon; for as he wore a coat of a fine white rein deer skin, it was all over besmeared with soot; hereupon Malina, or the Sun, went out to light a bit of moss; Anningait, or the Moon, did the same, but the flame of his moss was extinguished; this makes the Moon look like a fiery coal, and not shine so bright as the Sun. The Moon then run after the Sun round about the house to catch her; but she, to get rid of him, flew up into the air, and the Moon pursuing her, did likewise; and thus they still continue to pursue one another, though the Sun's career is much above that of the Moon.[42]

They also tell us, that the Moon is yet obliged to seek for his livelihood upon the earth and sea, in catching of seals, as a food he formerly was used to; which they pretend he is doing, when he appears not in the air: nay, they do not stick to say, that she now and then comes down to give their wives a visit, and caress them; for which reason no woman dare sleep lying upon her back, without she first spits upon her fingers and rubs her belly with it.

For the same reason the young maids are afraid to stare long at the moon, imagining they may get a child by the bargain. During the eclipse of the sun no man dare stir out of the house; and likewise when the moon is eclipsed, no woman goes abroad, because they fancy that both hate the sex of the other. The sun for joy puts on her pendants, or ear-bobs; the reason of which they take to be the hatred she bears against her brother, which also reaches to his sex. As on the contrary, the Greenland women wear their pendants at the birth of a boy, because so useful a creature is come into the world. Their notion about the stars is, that some of them have been men, and other different sorts of animals and fishes. The faint light of some stars they attribute to their eating the kidney; and brightness of others to their feeding upon liver. They give also names to many stars and constellations, *viz.* the three stars in the belt of Orion, they name Siektut,

---

[42] They assign the Moon a house in the Western part of the world, where he is often visited and resorted to by the angekkuts. And the Sun, they say, has her abode in the East; but she is inaccessible on account of her heat, which keeps the angekkuts at a distance; at which she is sorely grieved, because she cannot learn by them how matters stand upon Earth.

that is separated; because these three, they say, before their metempsychosis, or rather metamorphosis, were three honest Greenlanders, who being out at sea, a seal catching, were bewildered, and not being able to find the shore again, were taken up into Heaven.

Ursa Major, the great bear star, is styled by those that dwell in 64°, Tugto, or rein deer; while they that live in the bay of Disco at 69°, call it Asselluit, the name of a tree, to which they tie their line when they shoot seals. Taurus, the second sign in the Zodiac, is named Kellukturset, or kennel of hounds, who seem to have a bear among them; by this constellation they reckon their hours by night. Iversuk, that is, two persons that contend with songs or verses in taunting one another, as is customary among the Greenlanders. These two stars are in the constellation Taurus, of which heretofore, Aldebaran or Nennerroak, that is, a light which lights the two singers. Canis Major is called Nelleraglek, which is the name of a man amongst them; this they say has got on a coat of rein deer's skin. Gemini, Auriga, and Capella, are named Killaub Kuttuk, that is, the breast bone of Heaven.

When two stars seem to meet together, they say, that they are visiting one another; others will have it to be two women, who being rivals, take one another by the hair.

Concerning thunder and lightning, they say that two old women live together in one house in the air, who now and then fall out and quarrel about a thick and stiff outstretched seal skin (because such a skin, if beaten as a drum, has some likeness to the noise of thunder); while they are thus by the ears together, down comes the house with great bouncing and cracking, and the lamps are broken, the fires and broken pieces fly about in the air, and this, in their philosophy, is thunder and lightning.

In their astronomical system, the heaven turnabout upon the point of a huge rock. The snow, according to their fancy, is the blood of the dead, on account that it turns reddish if you keep it in the mouth. The rain comes from a ditch or wear above in Heaven; when it overflows there, it rains here below.

They have no calendar or almanacks, nor do they compute or measure the time by weeks or years, but only by months; beginning their computation from the Sun's first rising above their horizon in the winter; from whence they tell the month, to know exactly the season, in which every sort of fishes, sea animals, or birds seek the land; according to which they order their business.

As nonsensical now as these notions of the Greenlanders are (as they in reality are), yet they come short of the Egyptian King Ptolemy's infatuation, who by the loathsome flattery of his astronomers was persuaded that his Queen Berenice's head of hair was translated into Heaven and astrified, if I may say so; which constellation to this day goes by the name of Coma Be renices, or Berenice's hair; and what travellers relate of China and the East Indies, where some are of opinion, that the Sun's eclipse is nothing but that a certain devil or sprite sometimes swallows up the Sun, and then again spews it out.

Chapter 20

THE CAPACITY OF THE GREENLANDERS, AND THEIR INCLINATION TOWARDS THE KNOWLEDGE OF GOD, AND THE CHRISTIAN RELIGION; AND BY WHAT MEANS THIS MAY EASILY BE BROUGHT ABOUT

As the Greenlanders are naturally very stupid and indolent; so are they likewise very little disposed to comprehend and consider the divine truths which we expound to them; and notwithstanding people in years seem to approve of the Christian doctrine, yet it is with a surprising indifference and coldness. For they can neither comprehend the miserable condition they are in; nor do they rightly understand and value the exceeding great mercy and loving kindness God has shown towards mankind in his dear Son Christ Jesus, so as to move them to any desire and longing after it; some few excepted. This is to me an undeniable evidence that the carnally-minded man cannot comprehend the things that belong to God; for to him they seem to be foolish, and he cannot know them, as the Apostle speaks, *1 Cor. ii.* But as they in general are so credulous, that one can make them believe anything, so they are likewise in this grand affair. They never question what they are taught of God and Christ; but at the same time it never takes any rooting in their mind, because it passes without any consideration and feeling. For which reason they do not contradict or dispute with us the matters proposed; and very few have offered any objections, or desired any difficulty to be explained. And as their behaviour is silly and childish, so we have used the same method in teaching them, as we do to instruct little children; inculcating the Christian truths into their mind by frequent repetitions, and making use of simple and obvious comparisons, which, I thank God Almighty, has not wanted his blessing. For I have perceived in some the working of his grace in a serious amendment of their lives; and their endeavours have been to advance in the way to perfection, though all as yet is but a beginning and infancy, as we have mentioned in the last year's Memoirs or Relations of Greenland.

It is a matter which cannot be questioned, that if you will make a Christian out of a mere savage and wild man, you must first make him a reasonable man, and the next step will be easier. This is authorised and confirmed by our Saviour's own method. He makes a beginning from the earthly things; he proposes the mysteries of the kingdom of God in parables and similitudes. The first care taken in the conversion of Heathens is to remove out of the way all obstacles which may hinder their conversion, and render them unfit to receive the Christian doctrine, before anything successfully can be undertaken in their behalf.

It would contribute a great deal to forward their conversion, if they could by degrees be brought into a settled way of life, and to abandon this sauntering and wandering about from place to place to seek their livelihood. But this cannot be hoped until a Christian nation comes to be settled among them (I mean in such places where the ground is fit for tillage and pasturage) to teach them, and by little and little accustom them to a quiet and more useful way of life, than that which they now follow.

They should also be kept under some discipline, and restrained from their foolish superstitions, and from the silly tricks and wicked impostures of their angekkuts, which ought to be altogether prohibited and punished.

Yet my meaning is, not that they, by force and constraint, should be compelled to embrace our religion, but to use gentle methods. Is it not allowed in the church of Christ to make use of Christian discipline at times and seasons, with prudence and due moderation; which is a powerful means to advance the growth of piety and devotion? How much more is it necessary to apply the same means here to grub up an untilled ground, where a new church is to be planted? Else it would be the same imprudence as to throw good seed into thorns and briars, which would choak the seed.

But as the chief fruit of our labours and teaching is to be expected from the growing youth, so if some good regulations and small foundations were laid for the bringing up a number of children in the Christian faith and piety, no doubt God would prosper it; inasmuch as these poor children and growing youth are very tractable and teachable, and good natured; showing no inclination or propensity to vice. Neither do they want capacity; for I have found they will take anything as soon as any of our own children. Now if these gifts or natural talents were forwarded by the gifts of grace, who would question their growth and advancement in the Christian faith and virtues, which would ripen to the full harvest of eternal happiness? Good God! how easy a thing would it be to help these poor wretches out of their misery, if those that God has blessed with wealth were heavenly minded, and would be sensible of the wretched condition of their fellow creatures, and contribute out of their abundance to the founding of a school in these parts, and the providing of other most necessary things!

His Majesty, out of his wonted most glorious zeal for the growth and advancement of the church of Christ, has most graciously provided, by a considerable sum of money yearly set apart, for the Greenland Missionaries' entertainment, which royal bounty continues to this day; for which goodness the most gracious God will bless his Majesty and all the royal hereditary house, and be their reward for ever. But as a good deal of this bounty money must be employed in the promoting of trade (without which the mission could not subsist), but little remains for promoting the proper end of the mission, which is the conversion of the Heathens, in which at present are employed no more than four missionaries, and two catechists, besides some few charity children belonging to both colonies, whose entertainment is to be provided for. Hitherto we have not been able to do great matters, but contented ourselves with some excursions here and there instructing the natives; who likewise, when they have had an opportunity, come to us with their families to be instructed. But as these excursions of ours, and those visits of theirs have not been very frequent, and only for a short time, by reason of the impossibility of travelling at all seasons, which has obliged us to leave them for a while to deal for themselves; it is not to be expected that our pains-taking should have had that success, which would attend it, if there were missionaries settled in different stations amongst them. For in several years we count but between twenty and thirty aged persons, and a hundred and odd young ones, that have been found capable to receive the holy sacrament of baptism. If amongst ourselves we had no schools, nor other pious foundations, for the instruction and Christian education of youth and old people, pray what great feats would one or two teachers in a whole country be able to do, by once or twice a year taking a journey throughout the land, and preaching a passage sermon? The apostles of Christ did not think this method sufficient; but after they had preached the word of God up and down, they besides ordained and constituted teachers and catechists everywhere. And if so wholesome a method be followed in Greenland, who will question a happier success?

And this is all I at present have to say of the affairs of Greenland; leaving it to the judgment of others to be made out and decided, whether Greenland is a country that deserves to be improved and taken care of, or no? And whether its inhabitants may be called happy, or no? All things well pondered, both the affirmative and negative may be true, without the least contradiction. For Greenland can pass for no better than a dismal and pitiful country, in regard to the greatest part of it, *viz.* all the inland country, which is perpetually covered with ice and snow, that never melts, and therefore of no use to mankind; and as to the remaining part, on the sea side, most of it lies uncultivated and uninhabited. But here it may again be said, that as to the first part, or the inland country, it is a thing that is past remedy; but as to the last part, or the sea side, it may be put in a better state by settlements, and manuring, so that it may recover its former fertility; and thus it might be reckoned a good and profitable country, provided the formerly inhabited tracks of land were anew settled and peopled. I will forbear to mention the great wealth and richness, which lies hidden in the Greenland seas, and can never be exhausted.

From the land I will go to the inhabitants, which everybody will think more wretched than happy, considered as destitute of the true knowledge of their Creator; and besides lead but very poor and despicable lives. The knowledge of God is undoubtedly that which affords the greatest happiness to mankind; as the want of it makes one the most wretched of all beings. But who would dare to deny it, if I should find out somebody yet more wretched than they? And such there are who have been blessed with the true knowledge of God; yet do nevertheless refuse him that obedience, which, as our Creator and Master, and in regard of our redemption and a thousand other particular kindnesses, he has the best of titles to demand it upon, according as he requires it of us in his holy Word. If the life of the Greenlanders, which we call poor and despicable, with respect to morality, be compared to that of the most pretended Christians; I am afraid they will confound others on the great Day of Judgment. For though they have no law, yet by the light of nature do some of the works of the law, as the apostle says, Rom. ii. What thoughts will anyone harbour, who seriously considers the predominant passions, as greediness after gain, covetousness, unmeasured ambition and pride, sumptuous, voluptuous, and prodigal lives; envy, hatred, and mutual persecutions, and innumerable other vices and crimes of most Christians? Can anyone help thinking, but that such evil doers (the remotest from the life, which is God alone) must be deemed the most unhappy of all? Whilst on the other hand, the Greenlanders pass their lives, as I may say, in a natural innocence and simplicity. Their desires do not extend farther than to necessary things; pomp and pride is unknown to them; hatred, envy, and persecution never plagued them; neither do they affect the dominion over one another. In short, everyone is contented with his own state and condition, and are not tormented with unnecessary cares. Is not this the greatest happiness of this life? O happy people! what better things can one wish you, than what you already possess? Have you no riches? yet poverty does not trouble you. Have you no superfluity? yet you suffer no want. Is there no pomp and pride to be seen among you? neither is there any slight or scorn to be met with. Is there no nobility or high rank amongst them? neither is there any slavery or bondage. What is sweeter than liberty? And what is happier than contentedness? But one thing is yet wanting: I mean, the saving knowledge of God and his dear son Christ Jesus, in which alone consists eternal life and happiness. John xvii. And this is what we offer you, in preaching to you the holy Gospel.

Now, God, who bade light shine forth in darkness, enlighten your hearts, in the light of the knowledge of God's glorious appearance in and through Christ Jesus. May he deliver your souls from the slavery of the Devil, and of sinful lusts, as you are free from corporeal bondage, to the end that you always may be free with the Lord both in soul and body. Amen.

The End.

Charles Wood,
Printer, Poppin's Court, Fleet Street, London.

INDEX

CHANGES IN THE ARCTIC: ISSUES AND CHALLENGES

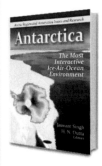

EDITOR: Julian C. Stephenson

SERIES: Arctic Region and Antarctica Issues and Research

BOOK DESCRIPTION: The Congressional Research Service (CRS) reports and Environmental Protection Agency Report (EPA) included in this book provide the readers with the issues, challenges and effects we face today due to Arctic climate changes.

HARDCOVER ISBN: 978-1-53613-823-8
RETAIL PRICE: $160

ANTARCTICA: THE MOST INTERACTIVE ICE-AIR-OCEAN ENVIRONMENT

EDITORS: Jaswant Singh (Avadh University, Faizabad U.P., India), H.N. Dutta (Roorkee Engineering & Management Technology Institute, Shamli, India)

SERIES: Arctic Region and Antarctica Issues and Research

BOOK DESCRIPTION: This book fills the gaps in the process of understanding Antarctic science.

HARDCOVER ISBN: 978-1-61122-815-1
RETAIL PRICE: $245